Materials for Engineering

W. Bolton

Butterworth-Heinemann Ltd
Linacre House, Jordan Hill, Oxford OX2 8DP

℟ A member of the Reed Elsevier group

OXFORD   LONDON   BOSTON
MUNICH   NEW DELHI   SINGAPORE   SYDNEY
TOKYO   TORONTO   WELLINGTON

First published 1994

© Butterworth-Heinemann Ltd 1994

British Library Cataloguing in Publication Data
Bolton , W.
   Materials for Engineering
   I. Title
   620.11

ISBN 0 7506 1838 8

Library of Congress Cataloguing in Publication Data
Bolton, W. (William), 1933–
   Materials for engineering/W. Bolton.
   p. cm.
   Includes index.
   ISBN 0 7506 1838 8
   1. Materials. I. Title.
   TA403.B65   93–51083
   620. 1'1–dc20   CIP

Composition by Scribe Design, Gillingham, Kent
Printed and bound in Great Britain by Bath Press

Contents

Contents

Preface

This book has been designed to comprehensively cover the GNVQ Engineering Mandatory unit at Advanced level of *Engineering Materials*, the Manufacturing Optional units at Intermediate and Advanced levels (formerly levels 2 and 3) of *Select and Test Materials* (U1016547) and *Select Materials* (U1016553). In addition, it covers the BTEC unit *Materials for Engineering* (171J). It is also seen as being of relevance to other courses where engineers require an introduction to the selection of materials.

The rationale is to introduce engineering students to:

• the properties and testing of materials used in manufacturing,
• the recognition of a relationship between the properties and the microstructure of the materials,
• the recognition of how the properties may be changed through modifications in composition, structure and processing,
• the selection of materials for particular applications,
• a knowledge of the requirements for safe procedures.

The book stems from the author's previous books on earlier BTEC units, namely *Engineering Materials 2, Engineering Materials 3, Materials Technology for Technicians 2* and *Materials Technology for Technicians 3*. It is, however, a complete rewrite. The book includes case studies, worked examples and a large number of problems, with answers supplied for all.

W. Bolton

<table>
<tr><td>**1**</td><td># Properties of materials</td></tr>
</table>

Outcomes At the end of this chapter you should be able to:

- Recognize the link between the selection of materials for a product and the properties required of them by the product.
- Communicate in appropriate technical terms about the properties of materials.
- Recognize the properties characteristic of different groups of materials.

1.1 Introduction to materials selection

What materials could be used for containers of Coca-Cola? Well, you can buy Coca-Cola in aluminium cans, in glass bottles and in plastic bottles. What makes these materials suitable and others not? In order to attempt to answer this question we need to discuss the properties of materials. Thus we might talk about the need for the container material to be:

1. Rigid, so that the container does not stretch unduly, i.e. become floppy, under the weight of the Coca-Cola.
2. Strong, so that the container can stand the weight of the Coca-Cola without breaking.
3. Resistant to chemical attack by the Coca-Cola.
4. Able to keep the 'fizz' in the Coca-Cola, i.e. not to allow the gas to escape through the walls of the container.
5. Low density so that the container is not too heavy.
6. Cheap.
7. Easy and cheap to process to produce the required shape.

You can no doubt think of more requirements. The selection of a material thus involves balancing a number of different specifications and making a choice of the material which fulfils as many as possible as well as possible.

Consider another product, a bridge. What are the requirements for the material to be used in a bridge? These are likely to include:

1   Strength so that when the bridge is subject to loads such as people, cars, lorries, etc. crossing it, then it will not break.
2   Stiff enough so that the bridge will not stretch unduly under the load.
3   Can be produced and joined in lengths long enough to span the gap to be bridged.
4   The materials costs and the fabrication costs are not too high.
5   Resists or can be protected from atmospheric corrosion.
6   Can be maintained at a reasonable cost over a period of years.

You can no doubt add more requirements. Materials which are used are wood, steel and reinforced concrete.

The selection of a material depends on the properties required of it so that it can fulfil the uses required of it.

1.1.1   The requirements of materials

The selection of a material from which a product can be manufactured depends on a number of factors. These are often grouped under three main headings, namely:

1   The requirements imposed by the conditions under which the product is used, i.e. the service requirements. Thus, if a product is to be subject to forces then it might need strength, if subject to a corrosive environment then it might require corrosion resistance.
2   The requirements imposed by the methods proposed for the manufacture of the product.. For example, if a material has to be bent as part of its processing, it must be ductile enough to be bent without breaking. A brittle material could not be used.
3   Cost.

1.2   Properties of materials

Materials selection for a product is based upon a consideration of the properties required. These include:

1   Mechanical properties. These are displayed when a force is applied to a material and include strength, stiffness, hardness, toughness and ductility.
2   Electrical properties. These are seen when the material is used in electrical circuits or components and include resistivity, conductivity and resistance to electrical breakdown.
3   Magnetic properties. These are relevant when the material is used as, for example, a magnet or part of an electrical component such as an inductor which relies on such properties.
4   Thermal properties. These are displayed when there is a heat input to a material and include expansivity and heat capacity.

5   Physical properties. These are the properties which are characteristic of a material and determined by its nature, including density, colour, surface texture.
6   Chemical properties. These are, for example, relevant in considerations of corrosion and solvent resistance.

The properties of materials are often changed markedly by the treatments they undergo. Thus, for example, steels can have their properties changed by heat treatment, such as annealing, which involves heating to some temperature and slowly cooling or quenching, i.e. heating and then immersing the material in cold water. They can also have their properties changed by working. For example, if you take a piece of carbon steel and permanently deform it then it will have different mechanical properties from those existing before that deformation (see Chapter 4 for more information).

In the following, some of the key properties listed above are discussed and the quantities which are used as a measure of them defined. Appendix 2 includes these, together with other properties and terms, in an alphabetical listing. Appendix 1 is a discussion of units and unit prefixes.

1.3   Mechanical properties

The mechanical properties are about the behaviour of materials when subject to forces. When a material is subject to external forces, then internal forces are set up in the material which oppose the external forces. The material can be considered to be rather like a spring. A spring, when stretched by external forces, sets up internal opposing forces which are readily apparent when the spring is released and they force it to contract. When a material is subject to external forces which stretch it then it is said to be in *tension* (Figure 1.1(a)). When a material is subject to forces which squeeze it then it is said to be in *compression* (Figure 1.1(b)). If a material is subject to forces which cause it to twist or one face to slide relative to an opposite face then it is said to be in *shear* (Figure 1.1(c)).

In discussing the application of forces to materials an important aspect is often not so much the size of the force as the force applied per unit area. Thus, for example, if we stretch a strip of material by a force F applied over its cross-sectional area A, then the force applied per unit area is F/A. The term *stress* is used for the force per unit area.

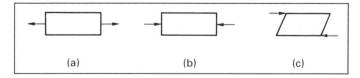

(a)   (b)   (c)

Figure 1.1   *(a) Tensile; (b) compressive; (c) shear forces*

$$\text{Stress} = \frac{\text{force}}{\text{area}}$$

Stress has the units of pascal (Pa), with 1 Pa being a force of 1 newton per square metre, i.e. $1 \text{ Pa} = 1 \text{ N/m}^2$ (see Appendix 1 on units). The stress is said to be *direct stress* when the area being stressed is at right angles to the line of action of the external forces, as when the material is in tension or compression. Shear stresses are not direct stresses since the forces being applied are in the same plane as the area being stressed. The area used in calculations of the stress is generally the original area that existed before the application of the forces. The stress is thus sometimes referred to as the *engineering stress*, the term *true stress* being used for the force divided by the actual area existing in the stressed state.

When a material is subject to tensile or compressive forces it changes in length. The term *strain* is used for

$$\text{Strain} = \frac{\text{change in length}}{\text{original length}}$$

Since strain is a ratio of two lengths it has no units. Thus we might, for example, have a strain of 0.01. This would indicate that the change in length is $0.01 \times$ the original length. However, strain is frequently expressed as a percentage.

$$\text{Strain as a \%} = \frac{\text{change in length}}{\text{original length}} \times 100$$

Thus the strain of 0.01 as a percentage is 1%, i.e. this is when the change in length is 1% of the original length.

Example
A bar of material with a cross-sectional area of 50 mm² is subject to tensile forces of 100 N. What is the tensile stress?

The tensile stress is the force divided by the area and is thus

$$\text{Tensile stress} = \frac{100}{50} \text{ N/mm}^2$$

$$= \frac{100}{50 \times 10^{-6}} \text{ N/m}^2 \text{ or Pa}$$

$$= 2 \text{MPa}$$

Example
A strip of material has a length of 50 mm. When it is subject to tensile forces it increases in length by 0.020 mm. What is the strain?

The strain is the change in length divided by the original length and is thus

$$\text{Strain} = \frac{0.020}{50} = 0.000\ 04$$

Expressed as a percentage, the strain is

$$\text{Strain} = \frac{0.020}{50} \times 100 = 0.04\%$$

1.3.1 Strength

The *strength* of a material is the ability of it to resist the application of forces without breaking. The forces can be tensile, compressive or shear. The tensile strength is defined as the maximum tensile stress the material can withstand without breaking, i.e.

$$\text{Tensile strength} = \frac{\text{maximum tensile forces}}{\text{original cross-sectional area}}$$

The compressive strength and shear strength are defined in a similar way. The unit of strength is the pascal (Pa), with 1 Pa being 1 N/m². Strengths are often millions of pascals and so the MPa is often used, 1 MPa being 10^6 Pa or 1 million Pa.

Often it is not the strength of a material that is important in determining the situations in which a material can be used but the value of the stress at which the material begins to yield. If gradually increasing tensile forces are applied to, say, a strip of mild steel then initially when the forces are released the material springs back to its original shape. The material is said to be *elastic*. If measurements are made of the extension at different forces and a graph plotted, then the extension is found to be proportional to the force and the material is said to obey *Hooke's law*. However, when a particular level of force is reached the material stops springing back completely to its original shape and is then said to show some *plastic* behaviour. This point coincides with the point on a force-extension graph at which the graph stops being a straight line graph, the so-called *limit of proportionality*.

Figure 1.2 shows the type of force-extension graph which would be given by a sample of mild steel. The limit of proportionality is point A. Up to this point Hooke's law is obeyed and the material shows elastic behaviour, beyond it shows a mixture of elastic and plastic behaviour. Dividing the forces by the initial cross-sectional area of the sample and the extensions by the original length converts the force-extension data into a stress–strain graph, as in Figure 1.3. The stress at which the material starts to behave in a non-elastic manner is called the *elastic limit*. Generally at almost the same stress the material begins to stretch without any further increase in force and is said to have yielded. The term *yield stress* is used for the stress at which this occurs. For some materials, such as mild steel, there are two yield points, termed the upper and the lower yield points. A carbon steel typically might have a tensile strength of 600 MPa and a yield stress of 300 MPa.

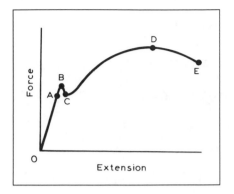

Figure 1.2 *Force-extension graph. A = limit of proportionality, B = upper yield point, C = lower yield point, D = maximum force, E = breaking point*

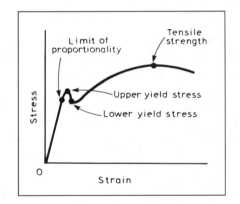

Figure 1.3 *Stress–strain graph*

In some materials, such as aluminium alloys, the yield stress is not so easily identified as with mild steel and the term *proof stress* is used as a measure of when yielding begins. This is the stress at which the material has departed from the straight-line force-extension relationship by some specified amount. The 0.1% proof stress is defined as that stress which results in a 0.1% offset, i.e. the stress given by a line drawn on the stress–strain graph parallel to the linear part of the graph and passing through the 0.1% strain value, as in Figure 1.4. A 0.2% proof stress is likewise defined as that stress which results in a 0.2% offset.

Stress–strain graphs are discussed in more detail in Chapter 3 when their determination and deductions that can be made from them are considered.

Example
A material has a yield stress of 200 MPa. What tensile forces will be needed to cause yielding with a bar of the material with a cross-sectional area of 100 mm^2 ?

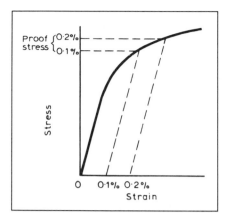

Figure 1.4   *Determination of proof stress*

Since stress is force/area then

> yield force = yield stress $\times$ area
>
> $= 200 \times 10^6 \times 100 \times 10^{-6} = 20\ 000$ N

Example
Samples are taken of cast aluminium alloys and give the following data. Which is the strongest in tension?

LM4 tensile strength 140 MPa
LM6 tensile strength 160 MPa
LM9 tensile strength 170 MPa

The strongest in tension is the one with the highest tensile strength and is LM9.

1.3.2   Stiffness

The *stiffness* of a material is the ability of a material to resist bending. When a strip of material is bent, one surface is stretched and the opposite face is compressed, as illustrated in Figure 1.5. The more a material bends, the greater is the amount by which the stretched surface extends and the compressed surface contracts. Thus a stiff material would be one that undergoes a small change in length when subject to such forces. This means a small strain when subject to such stress and so a small value of strain/stress, or conversely a large value of stress/strain. For most materials a graph of stress against strain gives initially a straight-line relationship, as illustrated in Figure 1.6. Thus a large value of stress/strain means a steep slope of the stress–strain graph. The quantity stress/strain when we are concerned with the straight-line part of the stress–strain graph is called the *modulus of elasticity* (or sometimes *Young's modulus*).

Materials for Engineering

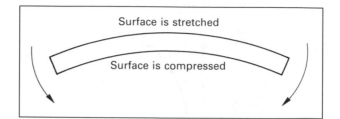

Figure 1.5   *Bending*

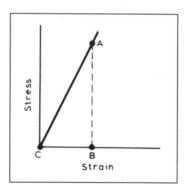

Figure 1.6   *Modulus of elasticity equals AB/BC*

$$\text{Modulus of elasticity} = \frac{\text{stress}}{\text{strain}}$$

The units of the modulus are the same as those of stress, since strain has no units. Engineering materials frequently have a modulus of the order of 1000 million Pa, i.e. 10^9 Pa. This is generally expressed as GPa, with 1 GPa = 10^9 Pa. Typical values are about 200 GPa for steels and about 70 GPa for aluminium alloys. A stiff material thus has a high modulus of elasticity. Thus steels are stiffer than aluminium alloys. For most engineering materials the modulus of elasticity is the same in tension as in compression.

Example
For a material with a tensile modulus of elasticity of 200 GPa, what strain will be produced by a stress of 4 MPa?
Since the modulus of elasticity is stress/strain then

$$\text{Strain} = \frac{\text{stress}}{\text{modulus}} = \frac{4 \times 10^6}{200 \times 10^9} = 0.000\ 02$$

Example
Which of the following plastics is the stiffest?

   ABS   tensile modulus 2.5 GPa
   Polycarbonate   tensile modulus 2.8 GPa

Polypropylene   tensile modulus 1.3 GPa
PVC   tensile modulus 3.1 GPa

The stiffest plastic is the one with the highest tensile modulus and therefore is the PVC.

1.3.3 Ductility/brittleness

If you drop a glass and it breaks then it is possible to stick all the pieces together again and restore the glass to its original shape. The glass is said to be a *brittle* material. If a car is involved in a collision, the bodywork is less likely to shatter like the glass but more likely to show permanent deformation, i.e. the material has shown plastic deformation (see Section 1.3.1). The term *permanent deformation* is used to changes in dimensions which are not removed when the forces applied to the material are taken away. Materials which develop significant permanent deformation before they break are called *ductile*. The car bodywork is likely to be mild steel, such a material being ductile. Figure 1.7 shows the types of stress–strain graphs given by brittle and ductile materials, the ductile one indicating a considerable extent of plastic behaviour.

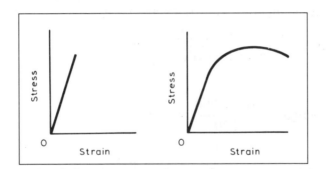

Figure 1.7   *Stress–strain graphs to fracture: (a) brittle materials; (b) ductile materials*

A measure of the ductility of a material is obtained by determining the length of a test piece of the material, then stretching it until it breaks and then, by putting the pieces together, measuring the final length of the test piece, as illustrated in Figure 1.8. A brittle material will show little change in length from that of the original test piece, but a ductile material will indicate a significant increase in length. The measure of the ductility is then the *percentage elongation*, i.e.

$$\% \text{ elongation} = \frac{\text{final} - \text{initial lengths}}{\text{initial length}} \times 100$$

A reasonably ductile material, such as mild steel, will have an elongation of about 20%, or more. A brittle material, such as a cast iron, will have an elongation of less than 1%.

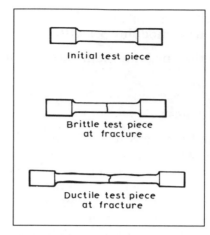

Figure 1.8 *Brittle and ductile test pieces after fracture*

Example

A material has an elongation of 10%. By how much longer will be a strip of the material of initial length 200 mm when it breaks?

The percentage elongation can be expressed as

$$\% \text{ elongation} = \frac{\text{change in length}}{\text{original length}} \times 100$$

Thus

$$\text{Change in length} = \frac{10 \times 200}{100} = 20 \text{ mm}$$

Example

Which of the following materials is the most ductile?

   80-20 brass % elongation 50%
   70-30 brass % elongation 70%
   60-40 brass % elongation 40%

The most ductile material is the one with the largest percentage elongation, i.e. the 70-30 brass.

Example

A sample of a carbon steel has a tensile strength of 400 MPa and an elongation of 35%. A sample of an aluminium–manganese alloy has a tensile strength of 140 MPa and an elongation of 10%. What does this tell you about the mechanical behaviour of the materials?

The higher value of the tensile strength of the carbon steel indicates that this material is stronger, and for the same cross-sectional area a bar of carbon steel could withstand higher tensile forces than a corresponding bar of the aluminium alloy. The higher percentage elongation of the carbon steel indicates that this material has a greater ductility than the aluminium alloy. Indeed, the value is such as to indicate that the carbon steel is very ductile.

1.3.4 Toughness

A tough material can be considered to be one that resists breaking. This we can take as meaning that a tough material requires more energy to break it than a less tough one. There are, however, a number of measures that are used for toughness.

Consider a length of material being stretched by tensile forces. When it is stretched by an amount x_1 as a result of a constant force F_1 then the work done is

Work = force × extension

$\text{Work} = F_1 x_1$

Thus if a force-extension graph is considered (Figure 1.9), the work done, when we consider a very small extension, is the area of that strip under the graph. The total work done in stretching a material to an extension x, i.e. through an extension which we can consider to be made up of a number of small extensions, is thus

$\text{Work} = F_1 x_1 + F_2 x_2 + F_3 x_3 + \ldots$

and so is the area under the graph up to x.

Since stress = force/area and strain = extension/length then

Work = (stress × area) × (strain × length)

Since the product of the area and length is the volume of the material, then

Work/volume = stress × strain

Thus the work done in stretching a material unit volume to a particular strain is the sum of the work involved in stretching the material to each of the strains up to this strain.

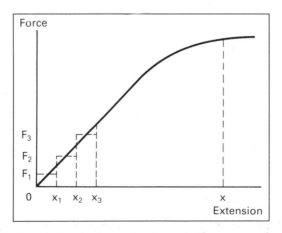

Figure 1.9 *Work done in producing extension* x *is the area under the graph up to* x

The area under a force-extension graph up to the breaking point is thus a measure of the energy required to break the material. The area under the stress–strain graph up to the breaking point is a measure of the energy required to break a unit volume of the material. A large area is given by a material with a large yield stress and high ductility (see Figure 1.7). Such materials can thus be considered to be tough.

An alternative way of considering toughness is the ability of a material to withstand shock loads. A measure of this ability to withstand suddenly applied forces is obtained by *impact tests*, such as the Charpy and Izod tests (see Chapter 3). In these tests a test piece is struck a sudden blow and the energy needed to break it is measured. The results are thus expressed in units of energy, i.e. joules (J). A brittle material will require less energy to break it than a ductile one. The results of such tests are often used as a measure of the brittleness of materials.

Another measure of toughness that can be used is fracture toughness. *Fracture toughness* can be defined as a measure of the ability of a material to resist the propagation of a crack. The toughness is determined by loading a sample of the material which contains a deliberately introduced crack of length $2c$ and recording the tensile stress σ at which the crack propagates. The fracture toughness, symbol K_c and usual units MPa m$^{1/2}$, is given by

$$K_c = \sigma \sqrt{\pi c}$$

The smaller the value of the toughness, the more readily a crack propagates. The value of the toughness depends on the thickness of the material, high values occurring for thin sheets and decreasing with increasing thickness to become almost constant in thick sheets. For this reason, a value called the *plane strain fracture toughness* K_{Ic} is often quoted. This is the value of the toughness that would be obtained with thick sheets. Typical values are of the order of 1 MPa m$^{1/2}$ for glass, which readily fractures when there is a crack present, to values of the order of 50 to 150 MPa m$^{1/2}$ for some steels and copper alloys. In such materials cracks do not readily propagate.

1.3.4 Hardness

The *hardness of a material* is a measure of its resistance to abrasion or indentation. A number of scales are used for hardness, depending on the method that has been used to measure it (see Chapter 3 for discussions of test methods). The hardness is roughly related to the tensile strength of a material, the tensile strength being roughly proportional to the hardness (see Chapter 3). Thus the higher the hardness of a material, the higher is likely to be the tensile strength.

1.4 Electrical properties

The electrical *resistivity* ρ is a measure of the electrical resistance of a material, being defined by the equation

$$\rho = \frac{RA}{L}$$

where R is the resistance of a length L of the material of cross-sectional area A. The unit of resistivity is the ohm metre. An electrical insulator such as a ceramic will have a very high resistivity, typically of the order of $10^{10}\,\Omega$ m or higher. An electrical conductor such as copper will have a very low resistivity, typically of the order of $10^{-8}\,\Omega$ m.

The electrical *conductance* of a length of material is the reciprocal of its resistance and has the unit of Ω^{-1}. This unit is given a special name, the siemen (S). The electrical *conductivity* σ is the reciprocal of the resistivity, i.e.

$$\sigma = \frac{1}{\sigma} = \frac{L}{RA}$$

The unit of conductivity is thus $\Omega^{-1}\,m^{-1}$ or $S\,m^{-1}$. Since conductivity is the reciprocal of the resistivity, an electrical insulator will have a very low conductivity, of the order of 10^{-10} S/m, while an electrical conductor will have a very high one, of the order of 10^{8} S/m.

The *dielectric strength* is a measure of the highest voltage that an insulating material can withstand without electrical breakdown. It is defined as

$$\text{Dielectric strength} = \frac{\text{breakdown voltage}}{\text{insulator thickness}}$$

The units of dielectric strength are volts per metre. Polythene has a dielectric strength of about 4×10^{7} V/m. This means that a 1 mm thickness of polythene will require a voltage of about 40 000 V across it before it will break down.

Example
An electrical capacitor is to be made with a sheet of polythene of thickness 0.1 mm between the capacitor plates. What is the greatest voltage that can be connected between the capacitor plates if there is not to be electrical breakdown? Take the dielectric strength to be 4×10^{7} V/m.

The dielectric strength is defined as the breakdown voltage divided by the insulator thickness, hence

$$\begin{aligned}
\text{breakdown voltage} &= \text{dielectric strength} \times \text{thickness} \\
&= 4 \times 10^{7} \times 0.1 \times 10^{-3} \\
&= 4000 \text{ V}
\end{aligned}$$

1.5 Thermal properties

The SI unit of temperature is the kelvin (K), with a temperature change of 1 K being the same as a change of 1°C.

The *linear expansivity* α or *coefficient of linear expansion* is a measure of the amount by which a length of material will expand when the temperature increases. It is defined as

$$\alpha = \frac{\text{change in length}}{\text{original length} \times \text{change in temperature}}$$

It has the unit of K^{-1}.

The *specific heat capacity* c is a measure of the amount of heat needed to raise the temperature of the material. It is defined as

$$c = \frac{\text{amount of heat}}{\text{mass} \times \text{change in temperature}}$$

It has the unit of $J\ kg^{-1}\ K^{-1}$. Weight-for-weight metals require less heat to reach a particular temperature than plastics. This is because metals have smaller specific heat capacities. For example, copper has a specific heat capacity of about $340\ J\ Kg^{-1}\ K^{-1}$ while polythene is about $1800\ J\ Kg^{-1}\ K^{-1}$.

The *thermal conductivity* λ of a material is a measure of its ability to conduct heat. There will only be a net flow of heat energy through a length of material when there is a difference in temperature between the ends of the material. Thus the thermal conductivity is defined in terms of the quantity of heat that will flow per second through a temperature gradient.

$$\lambda = \frac{\text{quantity of heat/second}}{\text{temperature gradient}}$$

It has the unit of $W\ m^{-1}\ K^{-1}$. A high thermal conductivity means a good conductor of heat. Metals tend to be good conductors. For example, copper has a thermal conductivity of about $400\ W\ m^{-1}\ K^{-1}$. Materials which are poor conductors of heat have low thermal conductivities. For example, plastics have thermal conductivities of the order of $0.03\ W\ m^{-1}\ K^{-1}$.

Example

A designer of domestic pans requires a material for a handle which would enable a hot pan to be picked up with comfort, the handle not getting hot. What quantity should he or she look for in tables in order to find a suitable material?

What is required is a material with a low thermal conductivity, probably a small fraction of a $W\ m^{-1}\ K^{-1}$.

1.6   Physical properties

The *density* ρ of a material is the mass per unit volume.

$$\rho = \frac{\text{mass}}{\text{volume}}$$

It has the unit of kg/m^3. It is often an important property that is required in addition to a mechanical property. Thus, for example, an aircraft under-carriage is required to be not only strong but also of low mass. Thus what is required is as high a strength as possible with as low a density as possible, i.e. a high value of strength/density. This quantity is often referred to as the *specific strength*. Steels tend to have specific strengths of the order of 50 to 100 MPa/Mg m^{-3} (note: 1 Mg is 10^6 g or 1000 kg), magnesium alloys about 140 MPa/Mkg m^{-3} and titanium alloys about 250 MPa/Mkg m^{-3}. Thus, for example, a lower-strength magnesium alloy would be preferred to a higher-strength, but higher-density, steel.

1.7 Chemical properties

Attack on materials by the environment in which they are situated is a major problem. The rusting of iron is an obvious example. Tables are often used giving the comparative resistance to attack of materials in various environments, e.g. in aerated water, in salt water, to strong acids, to strong alkalis, to organic solvents, to ultraviolet radiation. Thus, for example, in a salt water environment carbon steels are rated at having very poor resistance to attack, aluminium alloys good resistance and stainless steels excellent resistance.

1.8 The range of materials

Materials are usually classified into four main groups, these being metals, polymers and elastomers, ceramics and glasses, and composites. The following is a brief comparison, in general, of the properties of these main groups, Table 1.1 giving a comparison. Differences in the internal structure of the groups are discussed in Chapter 4.

1.8.1 Metals

Engineering metals are generally alloys. The term *alloy* is used for metallic materials formed by mixing two or more elements. For example, mild steel is an alloy of iron and carbon, stainless steel is an alloy of iron,

Table 1.1 *The range of properties*

Property	Metals	Polymers	Ceramics
Density (Mg m^{-3})	2-16	1-2	2-17
Melting point (°C)	200-3500	70-200	2000-4000
Thermal conductivity	High	Low	Medium
Thermal expansion	Medium	High	Low
Specific heat capacity	Low	Medium	High
Electrical conductivity	High	Very low	Very low
Tensile strength (MPa)	100-2500	30-300	10-400
Tensile modulus (GPa)	40-400	0.7-3.5	150-450
Hardness	Medium	Low	High
Resistance to corrosion	Medium–poor	Good–medium	Good

Note: 1 Mg m^{-3} = 1000 kg m^{-3}

chromium, carbon, manganese and possibly other elements. The reason for adding elements to the iron is to improve the iron's properties. Pure metals are very weak materials. The carbon improves the strength of the iron. The presence of the chromium in the stainless steel improves the corrosion resistance.

The properties of any metal are affected by the treatment it has received and the temperature at which it is being used. Thus heat treatment, working and interaction with the environment can all change the properties. In general, metals have high electrical and thermal conductivities, can be ductile and thus permit products to be made by being bent into shape, and have a relatively high modulus of elasticity and tensile strength.

1.8.2   Polymers and elastomers

Polymers can be classified as either *thermoplastics* or *thermosets*. Thermoplastics soften when heated and become hard again when the heat is removed. The term implies that the material becomes 'plastic' when heat is applied. Thermosets do not soften when heated, but char and decompose. Thus thermoplastic materials can be heated and bent to form required shapes, but thermosets cannot. Thermoplastic materials are generally flexible and relatively soft. Polythene is an example of a thermoplastic, being widely used in the forms of films or sheet for such items as bags, 'squeeze' bottles, and wire and cable insulation. Thermosets are rigid and hard. Phenol formaldehyde, known as Bakelite, is a thermoset. It is widely used for electrical plug casings, door knobs and handles.

The term *elastomers* is used for polymers which by their structure allow considerable extensions that are reversible. The material used to make rubber bands is an obvious example of such a material.

All thermoplastics, thermosets and elastomers have low electrical conductivity and low thermal conductivity, hence their use for electrical and thermal insulation. Compared with metals, they have lower densities and higher coefficients of expansion, are generally more corrosion resistant, have a lower modulus of elasticity, tensile strengths which are nearly as high as metals, are not as hard, and give larger elastic deflections. When loaded they tend to creep, i.e. the extension gradually changes with time. Their properties depend very much on the temperature so that a polymer which may be tough and flexible at room temperature may be brittle at $0°C$ and creep at a very high rate at $100°C$.

1.8.3   Ceramics and glasses

Ceramics and glasses tend to be brittle, have a relatively high modulus of elasticity, are stronger in compression than in tension, are hard, chemically inert, and have low electrical conductivity. Glass is just a particular form of ceramic, with ceramics being crystalline and glasses noncrystalline. Examples of ceramics and glasses abound in the home in the form of cups, plates and glasses. Alumina, silicon carbide, cement and concrete are examples of ceramics. Because of their hardness and abrasion resistance, ceramics are widely used for the cutting edges of tools.

1.8.4 Composites

Composites are materials composed of two different materials bonded together in such a way that one serves as the matrix and surrounds the fibres or particles of the other. There are thus composites involving glass fibres or particles in polymers, ceramic particles in metals (referred to as cermets) and steel rods in concrete (referred to as reinforced concrete). Wood is a natural composite consisting of tubes of cellulose in a natural polymer called lignin.

Composites are able to combine the good properties of other types of materials while avoiding some of their drawbacks. They can thus be made low density, with strength and a high modulus of elasticity. However, they generally tend to be more expensive to produce.

1.9 Costs

These can be considered in relation to the basic costs of the raw materials, the costs of manufacturing and the life and maintenance costs of the finished product.

Comparison of the basic costs of materials is often on the basis of the cost per unit weight or cost per unit volume. Thus, for example, if the cost of 10 kg of a metal is, say, £1 then the cost per kg is £0.1. If the metal has a density of 8000 kg/m³ then 10 kg will have a volume of 10/8000 = 0.001 25 m³ and so the cost per cubic metre is 1/0.001 25 = £800. Thus we can write

Cost per m³ = (cost/kg) × density

However, often a more important comparison is on the basis of the per unit strength or cost per unit stiffness for the same volume of material. This enables the cost of, say, a beam to be considered in terms of what it will cost to have a beam of a certain strength or stiffness. Hence if, for comparison purposes, we consider a beam of volume 1 m³ then if the tensile strength of the above material is 500 MPa, the cost per MPa of strength will be 800/500 = £1.60. Thus we can write, for the same volume,

$$\text{Cost per unit strength} = \frac{(\text{cost/m}^3)}{\text{strength}}$$

and similarly

$$\text{Cost per unit stiffness} = \frac{(\text{cost/m}^3)}{\text{modulus}}$$

The costs of manufacturing will depend on the processes used. Some processes require a large capital outlay and then can be employed to produce large numbers of the product at a relatively low cost per item. Others may have little in the way of setting-up costs but a large cost per unit product. See Chapter 5 for a discussion of processes.

The cost of maintaining a material during its life can often be a significant factor in the selection of materials. A feature common to many metals is the need for a surface coating to protect them from corrosion by the atmosphere. The rusting of steels is an obvious example of this and dictates the need for such activities as the continuous repainting of the Forth Railway Bridge.

Example

On the basis of the following data, compare the costs per unit strength of the two materials for the same volume of material.

Low-carbon steel: Cost per kg £0.10, density 7800 kg/m³, strength 1000 MPa

Aluminium alloy (Mn):   Cost per kg £0.22, density 2700 kg/m³, strength 200 MPa

For the steel, the volume of 1 kg is $1/7800 = 0.000\ 13$ m³ and so the cost per m³ is $0.1/0.00013 = £770$. The cost per MPa of strength is thus $770/1000 = £0.77$. For the aluminium alloy, the volume of 1 kg is $1/2700 = 0.000\ 37$ m³ and so the cost per m³ is $0.22/0.000\ 37 = £590$. Thus although the cost per kg is greater than that of the steel, because of the lower density the cost per cubic metre is less. The cost per MPa of strength is $590/200 = £2.95$. Hence on a comparison on the strengths of equal volumes, it is cheaper to use the steel.

Problems

1 What types of properties would be required for the following products?
   (a) A domestic kitchen sink.
   (b) A shelf on a bookcase.
   (c) A cup.
   (d) An electrical cable.
   (e) A coin.
   (f) A car axle.
   (g) The casing of a telephone.
2 For each of the products listed in problem 1, identify a material that is commonly used and explain why its properties justify its choice for that purpose.
3 Which properties of a material would you need to consider if you required materials which were:
   (a) stiff,
   (b) capable of being bent into a fixed shape,
   (c) capable of not fracturing when small cracks are present,
   (d) not easily broken,
   (e) acting as an electrical insulator,
   (f) a good conductor of heat,
   (g) capable of being used as the lining for a tank storing acid?
4 A colleague informs you that a material has a high tensile strength with a low percentage elongation. Explain how you would expect the material to behave.

5 A colleague informs you that a material has a high tensile modulus of elasticity and good fracture toughness. Explain how you would expect the material to behave.

6 What is the tensile stress acting on a strip of material of cross-sectional area 50 mm² when subject to tensile forces of 1000 N?

7 Tensile forces act on a rod of length 300 mm and cause it to extend by 2 mm. What is the strain?

8 An aluminium alloy has a tensile strength of 200 MPa. What force is needed to break a bar of this material with a cross-sectional area of 250 mm²?

9 A test piece of a material is measured as having a length of 100 mm before any forces are applied to it. After being subject to tensile forces it breaks and the broken pieces are found to have a combined length of 112 mm. What is the percentage elongation?

10 A material has a yield stress of 250 MPa. What tensile forces will be needed to cause yielding if the material has a cross-sectional area of 200 mm²?

11 A sample of high tensile brass is quoted as having a tensile strength of 480 MPa and an elongation of 20%. An aluminium–bronze is quoted as having a tensile strength of 600 MPa and an elongation of 25%. Explain the significance of these data in relation to the mechanical behaviour of the materials.

12 A grey cast iron is quoted as having a tensile strength of 150 MPa, a compressive strength of 600 MPa and an elongation of 0.6%. Explain the significance of the data in relation to the mechanical behaviour of the material.

13 A sample of a carbon steel is found to have an impact energy of 120 J at temperatures above 0°C and 5 J below it. What is the significance of these data?

14 Mild steel is quoted as having an electrical resistivity of 1.6×10^{-7}. Is it a good conductor of electricity?

15 A colleague states that he needs a material with a high electrical conductivity. Electrical resistivity tables for materials are available. What types of resistivity values would you suggest he looks for?

16 Aluminium has a resistivity of $2.5 \times 10^{-8} \, \Omega$ m. What will be the resistance of an aluminium wire with a length of 1 m and a cross-sectional area of 2 mm²?

17 How do the properties of thermoplastics differ from those of thermosets?

18 You read in a textbook that 'Designing with ceramics presents problems that do not occur with metals because of the almost complete absence of ductility with ceramics'. Explain the significance of the comment in relation to the exposure of ceramics to forces.

19 Compare the specific strengths, and costs per unit strength for equal volumes, for the materials giving the following data:

Low-carbon steel: Cost per kg £0.1, density 7800 kg/m³, strength 1000 MPa

Polypropylene: Cost per kg £0.2, density 900 kg/m³, strength 30 MPa

<table>
<tr><td>

2

</td><td>

Properties data

</td></tr>
</table>

Outcomes At the end of this chapter you should be able to:

- Recognize the role played by standards issued by national and international standards bodies in aiding effective communication between material producers, product manufacturers and consumers.
- Identify the range of sources from which information about the properties of materials can be obtained.
- In considering the suitability of a material for a particular application, determine what information is required and gather, collate, analyse and present it.

2.1 Standards

There are many thousands of standards laid down by national standards bodies such the British Standards Institution (BS) and international bodies such as the International Organization for Standardization (ISO). A standard is a technical specification drawn up with the cooperation and general approval of all interests affected by it with the aims of benefiting all concerned by ensuring consistency in quality, rationalizing processes and methods of operation, promoting economic production methods, providing a means of communication, protecting consumer interests, encouraging safe practices, and helping confidence in manufacturers and users. Thus, for example, there is a standard on the design of cylindrical and spherical pressure vessels subject to internal pressure (BS 5500). This standard covers the construction of pressure vessels for use in chemical plants and gives equations and software solutions by which the wall thickness can be calculated in order to ensure safe operation. There is a standard for the graphical symbols used in technical drawing (BS1553). This specifies the form symbols used in drawing should take so that all those reading the drawings will understand their significance.

Standards, both national and international, are used in the case of materials to ensure such things as consistency of quality and in the use of terms and rationalization of testing methods, and provide an efficient means of communication between interested parties. Thus if a material is stated by its producer to be to a certain standard, tested by the methods laid down by certain standards, then a customer need not have all the written details of what properties and tests have been carried out by the producer in order to know what properties to expect of the material. There is, for example, the standard for tensile testing of metals (BSEN 10002, a European standard adopted as the British Standard and replacing BS 18). This lays down such items as the sizes of the test pieces to be used (see Chapter 3 for details). There are standards for materials such as copper and copper alloy plate (BS 2875), steel plate, sheet and strip (BS 1449), the plastic polypropylene (BS 5139), etc. which lay down the composition and properties for such materials.

2.2 Data sources

Data on the properties of materials is available from a range of sources. These include:

1   Specifications issued by bodies responsible for standards, e.g. the British Standards Institution, the American Society for Metals, the International Organization for Standardization, etc. The standards operating in Britain are those issued by the British Standards Institution and listed in their catalogue. This lists all the British Standards and also whether they are also European and/or international standards.
2   Data books, e.g. the *ASM Metals Reference Book* (American Society for Metals, 1983), *Metals Reference Book* by R. J. Smithells (Butterworth, 1987), *Metals Databook* by C. Robb (The Institute of Metals, 1987), *Handbook of Plastics and Elastomers* edited by C. A. Harper (McGraw-Hill, 1975), *Newnes Engineering Materials Pocket Book* by W. Bolton (Heinemann-Newnes, 1989).
3   Computerized databases which give materials and their properties with means to rapidly access particular materials or find materials with particular properties, e.g. *Cambridge Materials Selector* (Cambridge University Engineering Department, 1992), *MAT.DB* (American Society for Metals, 1990).
4   Trade associations, e.g. the Copper Development Association which issues brochures giving technical details of compositions and properties of copper and copper alloy materials.
5   Data sheets supplied by suppliers of materials.
6   In-company tests. These are used to check samples of a bought-in material to ensure that it conforms to the standards specified by the supplier.

2.2.1 Coding systems

Coding systems are used to refer to particular metals. Such codes can relate to the chemical composition or the type of properties it has and are a concise way of specifying a particular material without having to write out its full chemical composition or properties. Thus, for example, steels might be referred to in terms of a code specified by the British Standards Institution. That for wrought steels is a six-symbol code. The first three digits of the code represent the type of steel:

000 to 199   Carbon and carbon–manganese types, the number being 100 times the manganese contents

200 to 240   Free-cutting steels, the second and third numbers being approximately 100 times the mean sulphur content

250   Silicon–manganese spring steels

300 to 499   Stainless and heat-resistant valve steels

500 to 999   Alloy steels with different groups of numbers within this range being allocated to different alloys. Thus 500 to 519 is when nickel is the main alloying element, 520 to 539 when chromium is.

The fourth symbol is a letter

A   The steel is supplied to a chemical composition determined by chemical analysis

H   The steel is supplied to a hardenability specification

M   The steel is supplied to mechanical property specifications

S   The steel is stainless

The fifth and sixth digits correspond to 100 times the mean percentage carbon content of the steel.

As an illustration of the above coding system, consider a steel with a code 070M20. The first three digits are 070 and since they are between 000 and 199 the steel is a carbon or carbon–manganese type. The 070 indicates that the steel has 0.70% manganese. The fourth symbol is M and so the steel is supplied to mechanical property specifications. The fifth and sixth digits are 20 and so the steel has 0.20% carbon.

Aluminium cast alloys use the code of LM followed by a number, the number being used to indicate a specific alloy listed in BS 1490. The coding system for wrought aluminium alloys is that of the Aluminium Association. This uses four digits with the first digit representing the principal alloying element, the second modifications to impurity limits, and the last two digits for the 1XXX alloys the aluminium content above 99.00% in hundredths and for others as identification of specific alloys. The 1XXX alloys have the principal alloying element of 99.00% minimum aluminium, the 2XXX copper as the principal alloying element, 3XXX manganese, 4XXX silicon, 5XXX magnesium, 6XXX magnesium and silicon, 7XXX zinc, 8XXX other elements.

The British Standards Institution coding for wrought copper and copper alloys consists of two letters followed by three digits. The two letters indicate the alloy group and the three digits the alloy within that group. Thus for copper and alloys containing a very high percentage of copper the letter used is C, for copper–aluminium alloys, i.e. aluminium bronzes, the letters are CA, for copper–beryllium alloys, i.e. beryllium bronzes, the letters are CB, for copper–nickel alloys, i.e. cupro-nickels, the letters are

CN, for copper–silicon alloys, i.e. silicon bronzes, the letters are CS, for copper–zinc alloys, i.e. brasses, the letters are CZ, for copper–zinc–nickel alloys, i.e nickel-silvers, the letters are NS, for copper–tin–phosphorus alloys, i.e. phosphor bronzes, the letters are PB. For cast copper alloys, letters followed by a digit are used. For example, AB1, AB2 and AB3 arc aluminium bronzes, LB1, LB2, LB4 and LB5 are leaded bronzes, SCB1 and SCB2 are general-purpose brasses.

The above gives an indication of some of the codes developed by standards bodies to assist communication between suppliers and users of materials. In addition, since the properties of materials depend on the heat treatment and degree of working they have undergone, there are codes to specify such treatments. The conditions are generally referred to as the *temper* of the material. Thus, for example, copper alloys are designated as temper O if the material is in the annealed condition, H if supplied in the hard condition resulting from cold working with ¼H and ½H to indicate quarter and half-hard conditions, M for as manufactured, W(H) for the heat treatment called solution treatment to the hardened condition. With aluminium alloys M indicates as manufactured, O annealed, H followed by a number between 1 and 8 the degree of hardening resulting from cold working, T followed by various letters different forms of heat treatment.

2.3 Data analysis

The procedure that might be adopted in searching for a material with the properties required for a particular product could be:

1 Identify the properties required.
2 Look in British or international standards, or data books or computer databases for materials with the required properties.
3 The above might refine the search to a particular material. Then, if further information is required, trade association publications or supplier data sheets can be consulted.

To illustrate the above, consider the search for a material for use as a conductor of electricity where high conductivity is the main property required. Since metals are, in general, good conductors and polymers and ceramics very poor ones then the choice would seem to be among metals. When tables are consulted the following information can be found for electrical conductivities, at 20°C:

Aluminium	40×10^6 S/m
Copper	64×10^6 S/m
Gold	50×10^6 S/m
Iron	11×10^6 S/m
Silver	67×10^6 S/m

If cost is also a factor then gold and silver are likely to be ruled out. Thus copper appears to be the optimum choice. If we now consider what form of copper then tables are likely to yield data in the following form:

C101	Electrolytic tough-pitch h.c. copper	101.5–100
C103	Oxygen-free h.c. copper	101.5–100
C105	Phosphorus deoxidized arsenical copper	50–35
C108	Copper–cadmium	92–80

The conductivities are not expressed in S/m but in units called IACS (international annealed copper standard) units and written as a percentage. This scale is based on 100% being the conductivity (or resistivity) of annealed copper at 20°C. This is a resistivity of $1.7241 \times 10^8 \, \Omega$ m or a conductivity of 58.00×10^6 S/m. Thus, if there are no other considerations C101 or C103 would appear to be the choice. Often, however, there are other factors to be taken into account, such as strength.

Consider another example, a requirement for a metal with a low melting point. The aim is to use the metals in die casting for the production of small components for toys, e.g. toy car steering wheels and drive shafts. The following are some of the melting points for metals that can be found from tables:

Aluminium	600°C
Lead	320°C
Magnesium	520°C
Zinc	380°C

Lead and zinc have the lowest melting points. If we add another requirement of reasonable strength in the as-cast condition then tables give:

| Lead | Tensile strength 20 MPa |
| Zinc | Tensile strength 280 MPa |

Thus, taking strength into account, zinc would appear to be the choice. There are, however, other considerations that have to be taken into account before a choice can be made. See Section 6.4.3.

Table 2.1 *Mechanical properties of cast irons*

Material	Tensile strength (MPa)	Yield stress (MPa)	% elongation
Grey irons			
BS 150	160	98	0.6
BS 180	180	117	0.5
BS 220	220	143	0.5
BS 260	260	170	0.4
BS 300	300	195	0.3
BS 350	350	228	0.3
BS 400	400	260	0.2
Malleable irons			
Blackheart B32-10	320	190	10
Blackheart B35-12	350	200	12
Whiteheart W38-12	380	200	12
Whiteheart W40-05	400	220	5
Whiteheart W45-07	450	260	7

Example

The data in Table 2.1 is taken from *Newnes Engineering Materials Pocket Book* by W. Bolton (Heinemann-Newnes, 1989) and gives the properties of a number of cast irons. Select, from those listed, a cast iron which combines high tensile strength with ductility.

A high ductility means a high percentage elongation. Grey irons have very low percentage elongations and so are brittle. The malleable irons show greater percentage elongations and thus the selection needs to be one of these irons. The malleable iron with the greatest tensile strength (Whiteheart W45-07) does not, however, have as high a percentage elongation as Whiteheart W38-12. Thus if ductility is more important than strength the choice might be Whiteheart W38-12.

Problems

1 Determine from tables the following information for materials at about 20°C:
   (a) the tensile strength of the carbon steel 1030 in the as-rolled condition,
   (b) the electrical conductivity on the IACS scale of the unalloyed aluminium 1060 in the annealed state,
   (c) the percentage elongation of the brass SCB1,
   (d) the yield stress of the manganese steel 120M19 in the quenched and tempered state,
   (e) the tensile strength of the stainless steel 302S31 in the soft state,
   (f) the density of the plastic ABS,
   (g) the plane strain fracture toughness of the plastic polypropylene,
   (h) the tensile strength of the elastomer natural rubber,
   (i) the thermal expansivity (linear coefficient of expansion) of the plastic high-density polythene,
   (j) the tensile modulus of the plastic ABS.

2 The rainwater guttering used for buildings is required to have a high stiffness per unit weight so that it does not sag under its own weight. Use tables to obtain the specific modulus or values of the modulus and density and hence compare cast iron, aluminium alloys and the plastic PVC as possible materials.

3 The panels used for car bodywork need to be in sheet form and stiff. Use tables to obtain modulus of elasticity values and hence compare carbon steel, an aluminium alloy, polypropylene and a composite formed by polyester with 65% glass fibre cloth.

4 The fan in a vacuum cleaner needs to be made of a low-density material and a high tensile strength, i.e. a high specific strength. The aluminium alloy LM6 has been suggested because the fan could then be die cast. Use tables to obtain the specific strength of the material.

5 The plastic ABS has been suggested for use as the casing for a radio. The properties required include high stiffness. Determine from tables the modulus of elasticity and compare it with other plastics.

6 The material high-tensile brass HTB1 has been suggested for use as a marine propeller. Use tables to obtain values of its tensile strength, 0.1% proof stress and percentage elongation.

7 The alloy steels 150M36 and 530M40 have been suggested as the materials for a car axle. Determine from tables the tensile strengths, yield stresses and percentage elongations for both materials, in the quenched and tempered state, so that a comparison can be made.

8 Table 2.2 is taken from *Newnes Engineering Materials Pocket Book* by W. Bolton (Heinemann-Newnes, 1989) and gives data for cast aluminium alloys when sand cast and in the as-manufactured condition. Select, from the list, a material which is likely to be tough.

9 Table 2.3 is taken from *Newnes Engineering Materials Pocket Book* and gives data for polymers. Select, from the list, a material which will be stiff and not too brittle.

10 Table 2.4 is taken from *Newnes Engineering Materials Pocket Book* and gives data for free-cutting steels which have been quenched and tempered to 550-660°C. Select, from the list, the strongest steel.

Table 2.2   *Mechanical properties of cast aluminium alloys*

Material	Tensile strength (MPa)	% elongation
LM4	140	2
LM5	140	3
LM6	160	5

Table 2.3   *Mechanical properties of polymers*

Polymer	Tensile strength (MPa)	Tensile modulus (MPa)	% elongation
ABS	17-58	1.4-3.1	10-140
Acrylic	50-70	2.7-3.5	5-8
Cellulose acetate	24-65	1.0-2.0	5-55
Cellulose acetate butyrate	18-48	0.5-1.4	40-90
Polyacetal, homopolymer	70	3.6	15-75
Polyamide, Nylon 6	75	1.1-3.1	60-320
Polyamide, Nylon 66	80	2.8-3.3	60-300

Table 2.4   *Mechanical properties of free-cutting steels*

Steel	Tensile strength (MPa)	Yield stress (MPa)	% elongation
212M36	550-700	340	20
216M36	550-700	380	15
220M44	700-850	450	15

<table>
<tr><td>

3

</td><td>

Materials testing

</td></tr>
</table>

Outcomes At the end of this chapter you should be able to:

*   Identify the standard tests used for key properties and materials.
*   Recognize the test procedures required.
*   Interpret test results explaining their validity and limitations.

Note: references are made for the various tests to British and European standards. However, for full details of tests the reader is referred to the relevant standards.

3.1 Standard tests

The standard tests used in Britain are those specified by the British Standards Association. These include:

BSEN 1002	Methods of tensile testing of metals (formerly BS18)
BS 131	Methods of notched bar tests
BSEN 10045	Charpy test
BS 240	Method for Brinell hardness test
BS 427	Method for Vickers hardness test
BS 891	Method for Rockwell hardness test
BS 1639	Method of bend testing of metals
BS 2782	Methods of testing plastics
BS 5714	Resistivity measurements for metals
BS 7448	Methods of test for plane strain fracture toughness of metallic materials.

Standards specified as BSEN are European standards which have been adopted as British standards. The above represents just a selection of the common tests. There are others which are not considered in this book, e.g. BS 3500 Methods for creep and rupture testing of metals, BS 3518 Methods of fatigue testing.

3.2 The tensile test

In a tensile test, measurements are made of the force required to extend a standard test piece at a constant rate, the elongation of a specified gauge length of the test piece being measured by some form of extensometer. British and European standards (BSEN 10002 Part 1) state that the rate at which the stresses are applied should be between 2 and 10 MPa/s if the tensile modulus is less than 150 GPa and between 6 and 30 MPa/s if the tensile modulus is equal to or greater than 150 GPa. In order to eliminate any variations in tensile test data due to differences in the shapes of test pieces, standard shapes and sizes are adopted.

3.2.1 The test piece

Test pieces are said to be *proportional* if the relationship between the gauge length L_0 and the cross-sectional area A of the gauge length is

$$L_0 = k \sqrt{A}$$

British and European standards specify that the constant k should have the value 5.65 and the gauge length should be 20 mm or greater. With circular cross-sections $A = \frac{1}{4}\pi d^2$ and thus to a reasonable approximation this value of k gives

$$L_0 = 5d$$

With circular cross-sectional areas which are too small for this value of k, a higher value may be used, preferably 11.3. When test pieces are proportional the same test results are given for the same test material when different size test pieces are used.

Figure 3.1 shows the standard size test pieces for round and flat samples of metals and Table 3.1 lists the standard dimensions that can be used. For the tensile test data for the same material to give essentially the same results, regardless of the length of the test piece used, it is vital that the standard dimensions are adhered to. An important feature of the dimensions is the radius given for the shoulders of the test pieces. Very small radii can cause localized stress concentrations which may result in the test piece failing prematurely.

3.2.2 Tensile test results

The results of tensile tests are obtained as force-extension data which are generally plotted, either manually or by a machine, as a force-extension graph. Since stress is force/original area and strain is extension/original gauge length then the graph is readily translated into stress–strain. From such graphs the following quantities can be determined (see Chapter 1 for discussion of the terms):

1. The *tensile strength*, being the stress corresponding to the maximum force.
2. The *yield stress*, being the stress at which the material begins to yield and show plastic deformation without any increase in load. The term

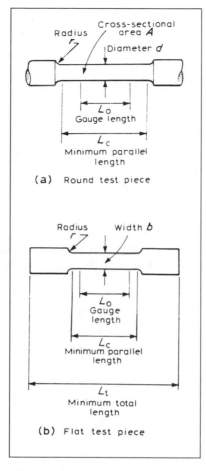

Figure 3.1   *Standard tensile test pieces*

Table 3.1   *Dimensions of standard test pieces*

Flat test pieces

b (mm)	L_0 (mm)	L_c (mm)	L_f (mm)
20	80	120	140
12.5	50	75	87.5

Round test pieces (proportional)

d (mm)	A (mm)^2	L_0 (mm)	L_c (mm)
20	314.2	100	110
10	78.5	50	55
5	19.6	25	28

Note: $k = 5.85$

upper yield stress is used for the value of the stress when the first decrease in force at the yield is observed and *lower yield stress* implicates the lowest value of stress during plastic yielding.

3 The *proof stress*, being the stress at which the non-proportional extension is equal to a specified percentage of the gauge length. 0.1% and 0.2% are the percentages commonly used.

4 The *tensile modulus*, being the slope of the stress–strain graph over its proportional region.

In addition, measurements of the gauge length before and after breaking enable the *percentage elongation* to be determined, this being the permanent elongation of the gauge length expressed as a percentage of the original gauge length.

Example
Figure 3.2 shows a stress–strain graph for a sample of mild steel. Determine (a) the upper yield stress; (b) the lower yield stress and (c) the tensile strength.

The upper yield stress is about 280 MPa, the lower yield stress about 240 MPa and the tensile strength about 400 MPa.

Example
Figure 3.3 shows part of the stress–strain graph for a sample of an aluminium alloy. Determine the 0.1% and 0.2% proof stresses.

The 0.1% proof stress is about 460 MPa and the 0.2% about 520 MPa.

3.2.3 Validity of tensile test data

The purpose of taking tensile test pieces and carrying out the tests is to obtain data which enable judgements to be made about the material from which the test piece was cut. The samples of a material have to be taken in such a way that the properties deduced from the tensile test are representative of the material as a whole. There may, however, be problems in assuming this. The following paragraphs outline some of these problems.

The properties of a product may not be the same in all parts of it. With a casting there may be different cooling rates in different parts of a casting, e.g. the surface compared with the core, or thin sections compared with thick sections. As a result, the internal structure of the material may differ and, as a consequence, the tensile properties differ. A tensile test piece cut from one part may not thus represent the properties of the entire casting. For the same reason, the properties of a separately cast test piece may not be the same as those of the cast product because the different sizes of the two lead to different cooling rates.

Note that if the mechanical properties of metals are looked up in tables you will often find that different values of the properties are quoted for different limiting ruling sections. The *limiting ruling section* is the maximum diameter of a round bar at the centre of which the specified properties

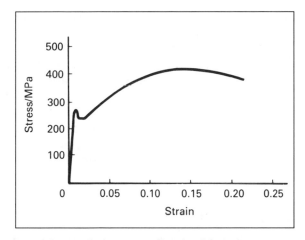

Figure 3.2   *Stress–strain graph for a sample of mild steel*

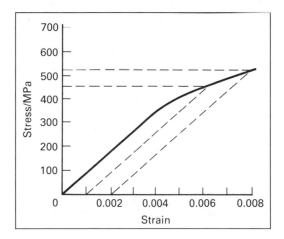

Figure 3.3   *Part of a stress–strain graph for an aluminium alloy*

may be obtained. The reason for the difference of mechanical properties of the same material for bars of different diameters is that during the heat treatment different rates of cooling occur at the centres of such bars due to their variations in sizes. Consequently there are differences in microstructure and hence in mechanical properties. For example, the steel 070M55 with a limiting ruling section of 19 mm may have tensile strengths of 850 to 1000 MPa, with 63 mm the strengths are 777 to 930 MPa and with 100 mm they are 700 to 830 MPa.

The properties of a product may not be the same in all directions. Thus, for example, with rolled sheet there is a directionality of properties with the tensile properties in the longitudinal, transverse and through-thickness directions of the sheet differing. Thus, for example, with rolled brass strip we might have tensile strengths of 740 MPa in the direction of the rolling and 850 MPa at right angles to it.

The temperature in service of the product may not be the same as that of the test piece when the tensile test data were obtained. The tensile properties of metals depend on temperature. In general, the tensile modulus and tensile strength both decrease with an increase in temperature and the percentage elongation tends to increase.

The rate of loading of a product may differ from that used with the test piece. The data obtained from a tensile test are affected by the rate at which the test piece is stretched, so in order to give a standardized result the tests are carried out at a constant stress rate, between 2 and 20 MPa/s if the tensile modulus is less than 150 GPa and between 6 and 30 MPa/s if it is equal to or greater than 150 GPa.

3.2.4 Interpreting tensile test data

The results from tensile tests can be used to determine the safe stresses to which a material can be subject. Thus, the higher the yield stress of a metal, the higher the stresses that it can be exposed to in service without yielding. Another important deduction that can be made is whether the material is brittle or ductile (see Chapter 1 and, in particular, Figure 1.7). A brittle material will show little plastic behaviour and have a low percentage elongation. A ductile material will show considerable plastic behaviour and have a high percentage elongation.

A consequence of the heat treatment and working of a material that occurs during the fabrication of products is a change in mechanical properties. Thus tensile test data enable the effectiveness of heat treatments and the effects of working to be monitored.

Example
Which of the materials shown in Figure 3.4 is (a) the most ductile; (b) the most brittle; (c) the strongest; (d) the stiffest?

(a) Material C, because it has the greatest plastic region to its graph. (b) Material B, because it has the least plastic region. (c) Material A, because it can experience the highest stress. (d) Material A, because it has the steepest slope for its proportional region.

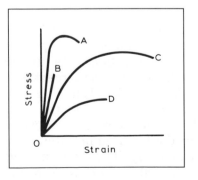

Figure 3.4 *Example*

3.2.5 Tensile tests for plastics

Tensile tests can be used with plastic test pieces to obtain stress–strain data. The term *tensile strength* has the same meaning as with metals. However, the tensile modulus, i.e. the slope of the stress–strain graph over the proportional region, cannot always be easily obtained. For many plastics there is no really straight-line part of the stress–strain graph. Thus, as a measure of the stiffness of the material, a modulus is defined in a different way. The *secant modulus* is obtained by dividing the stress at a strain of 0.2% by that strain, as illustrated in Figure 3.5.

The stress–strain properties of plastics are much more dependent than metals on the rate at which the strain is applied. Thus, for example, the tensile test may indicate a yield stress of 62 MPa when the rate of elongation is 12.5 mm/min but 74 MPa when it is 50 mm/min. Also the form of the stress–strain graph may change with a ductile material at low strain rates, becoming a brittle one at high strain rates. Figure 3.6 shows the general forms of stress–strain graphs for plastics at different strain rates. Another factor that is more marked than with metals is the effect of temperature on the properties of plastics.

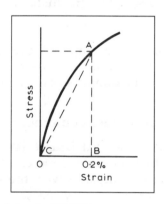

Figure 3.5   *The secant modulus is AB/BC*

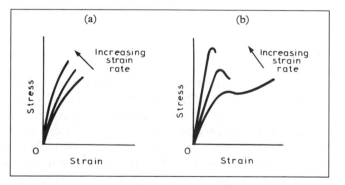

Figure 3.6   *The effect of strain rate on the stress–strain graphs for plastics. (a) A brittle plastic; (b) a ductile plastic*

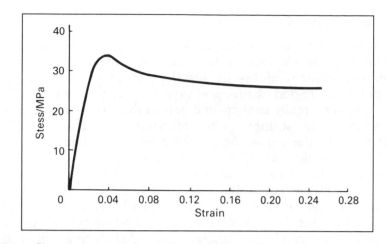

Figure 3.7   *Stress–strain graph for Novodur PK (adapted from Bayer UK Ltd)*

Example
Figure 3.7 is the stress–strain graph for a sample of ABS Novodur grade PK (courtesy of Bayer (UK) Ltd). Estimate (a) the tensile modulus and (b) the tensile strength.

(a) The tensile modulus is the slope of the proportional part of the stress–strain graph and is thus about 28/0.02 = 1400 MPa = 1.4 GPa. (b) The tensile strength is the maximum stress. This is about 34 MPa.

3.2.6   Verification of tensile test equipment

The British and European standard BSEN 10002 Part 2 describes how the force readings given by such a machine can be verified. A given force indicated by the machine is compared with the true force indicated by a force-proving instrument or exerted by masses. Three series of measurements should be taken with increasing force, each series having at least five steps at regular intervals from 20% of the maximum range of the scale.

3.3 Bend tests

A simple test that is often quoted by suppliers of materials as a measure of ductility is the bend test. This involves bending a sample of the material through some angle and determining whether the material is unbroken and free from cracks after the bending. There are a number of ways that can be used to carry out such a test, BS 1639 listing the British standards. The simplest method is the mandrell form of test shown in Figure 3.8(a), this being suitable for medium and thin thickness sheet for angles of bend up to 120°. Figure 3.8(b) shows how the test can be conducted on a vee block, this being suitable for medium thickness sheet

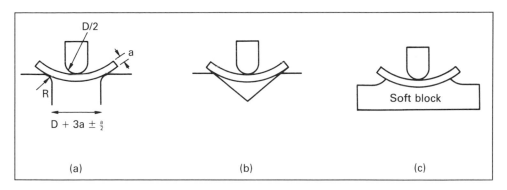

Figure 3.8 *Bend test. (a) Mandrel test; (b) bending on a vee block; (c) bending on a block of soft material*

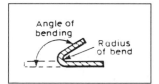

Figure 3.9 *The angle of bend*

with bend angles up to 90°. Figure 3.8(c) illustrates the form of test possible for thin sheet with bend angles up to 90°, the material being bent on a block of soft material. Other methods can also be used, e.g. bending round a mandrell, free bending and pressure bending (see the British Standard for more details). The results of a bend test are quoted in terms of the angle of bend that can be withstood without breaking or cracking, as illustrated in Figure 3.9.

3.4 Impact tests

Impact tests are designed to simulate the response of a material to a high rate of loading and involve a test piece being struck a sudden blow. There are two main forms of test: the *Izod* and *Charpy* tests. Both involve the same type of measurement but differ in the form of the test pieces. Both involve a pendulum swinging down from a specified height h_0 to strike the test piece (Figure 3.10). The height h to which the pendulum rises after striking and breaking the test piece is a measure of the energy used in the breaking. If no energy were used the pendulum would swing up to the same height h_0 as it started from, i.e. the potential energy mgh_0 at the top of the pendulum swing before and after the collision would be the same. The greater the energy used in the breaking, the lower the height to which the pendulum rises. If the pendulum swings up to a height h after breaking the test piece then the energy used to break it is $mgh_0 - mgh$.

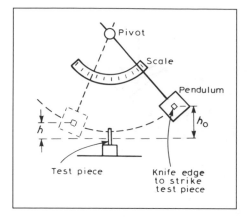

Figure 3.10   *The principle of the impact test*

3.4.1   Izod test pieces

With the Izod test the energy absorbed in breaking a cantilevered test piece is measured, as illustrated in Figure 3.11. The test piece has a notch and the blow is struck on the same face as the notch and at a fixed height above it.

In the case of metals, the test pieces used are generally either 10 mm square or 11.4 mm in diameter if round. Figure 3.12 shows details of one form of the square test piece. With the 70 mm length the notch is 28 mm from the top of the piece. If a longer length is used then more than one notch is used. With a length of 96 mm there are two notches on opposite faces, one 28 mm from the top and the other twice that distance from the top. With a longer length test piece of 126 mm there are three notches, on three of the faces. The first notch is 28 mm from the top, the second twice that distance and the third three times that distance from the top.

In the case of plastics, the test pieces are either 12.7 mm, square or 12.7 mm by 6.4 to 12.7 mm, depending on the thickness of the material concerned. Figure 3.13 shows details of such a test piece. With metals the pendulum strikes the test piece with a speed of between 3 and 4 m/s, and with plastics a lower speed of 2.44 m/s is used.

3.4.2   Charpy test pieces

With the Charpy test, the energy absorbed in breaking a test piece in the form of a beam is measured (Figure 3.14). The standard machine has the pendulum striking the test piece with an energy of 300 ± 10 J. The test piece is supported at each end and notched at the midpoint between the two supports. The notch is on the face directly opposite to where the pendulum strikes the test piece. The British and European Standard is BSEN 10045.

For metals, the test piece generally has a square cross-section of side 10 mm and length 55 mm with 40 mm between the supports. Figure 3.15 shows details of such a test piece and the forms of notch commonly used. With the V-notch reduced-width specimens of 7.5 mm and 5 mm can be used.

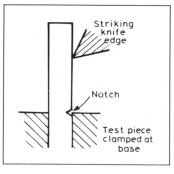

Figure 3.11 *Form of the Izod test (elevation view)*

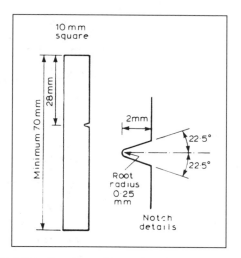

Figure 3.12 *Izod test piece for a metal*

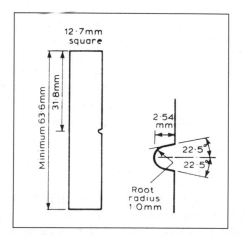

Figure 3.13 *Izod test piece for a plastic*

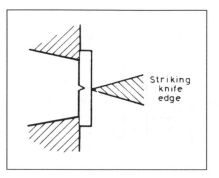

Figure 3.14 *Form of the Charpy test (plan view)*

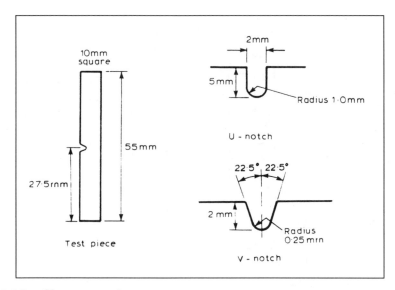

Figure 3.15 *Charpy test piece*

For plastics, the test pieces may be unnotched or notched. A standard test piece is 120 mm long, 15 mm wide and 10 mm thick in the case of moulded plastics. With sheet plastics the width can be the thickness of the sheet with a thickness 15 mm or the width between 5 and 10 mm with the thickness being 15 mm. The notch is U-shaped with a width of 2 mm and a radius of 0.2 mm at its base. For moulded plastics the depth below the notch is 6.7 mm, for the sheet plastics either 10 mm or two-thirds of the sheet thickness.

3.4.3 Impact test results

In stating the results of impact tests it is vital that the form of test is specified. There is no reliable relationship between the values obtained by the two forms of test and so values from one test cannot be compared with those from the other. In addition, there is no reliable relationship between the impact energies given for breaking test pieces of different sizes or different notches with the same test method. The impact energy is influenced by such factors as the temperature, the speed of impact, any degree of directionality in the properties of the material from which the test piece was cut, and the thickness of the test piece.

For both the Izod and Charpy tests, the impact strengths for metals are expressed in the form of the energy absorbed (for example, 30 J). For plastics, with the Izod test the results are expressed as the energy absorbed in breaking the test piece divided by the width of notch, and with the Charpy test as the energy absorbed divided by either the cross-sectional area of the specimen for unnotched test pieces or by the cross-sectional area behind the notch for notched test pieces (e.g. 2 kJ/m^2).

3.4.4   Interpreting impact test results

When a material is stretched, energy is stored in the material. Think of stretching a spring or a rubber band. When the stretching force is released the material springs back and the energy is released. However, if the material suffers a permanent deformation then all the energy is not released. The greater the amount of such plastic deformation, the greater the energy not released. Thus when a ductile material is broken, more energy is 'lost'.

The fracture of materials can be classified roughly as either brittle or ductile. With brittle fracture there is little plastic deformation prior to fracture and so little energy is required to break the test piece. With ductile fracture the fracture is preceded by a considerable amount of plastic deformation and so more energy is required to break the test piece. Thus the impact test can be used to give information about the type of fracture that occurs. For example, Figure 3.16 shows the effect of temperature on

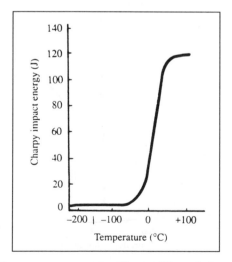

Figure 3.16   *Effect of temperature on the Charpy V-notch impact energies for a 0.2% carbon steel*

the Charpy V-notch impact energies obtained for test pieces of a 0.2% carbon steel. Above about 0°C the material gives ductile failures, below that temperature, brittle failures. Such graphs have a great bearing on the use that can be made of the material, since at low temperatures the steel can be easily shattered by impact.

The appearance of the fractured surfaces after an impact test also gives information about the type of fracture that has occurred. With a brittle fracture of metals, the surfaces are crystalline in appearance. With a ductile fracture, they are rough and fibrous. Also with ductile failure there is a significant reduction in the cross-sectional area of the test piece, but with brittle fracture there is virtually no such change. With plastics, a brittle failure gives fracture surfaces which are smooth and glassy or somewhat splintered, with a ductile failure the surfaces often have a whitened appearance. Also, with plastics, the change in cross-sectional area can be considerable with a ductile failure but negligible with brittle failure.

One use of impact tests is to determine whether heat treatment has been successfully carried out. A comparatively small change in heat treatment can lead to quite noticeable changes in impact test results. These can be more pronounced than changes in other mechanical properties such as percentage elongation or tensile strength. Figure 3.17 shows the effect of annealing to different temperatures on the Izod impact test results for cold-worked mild steel. The impact test can thus be used to indicate whether annealing has been carried out to the required temperature.

Example
A sample of unplasticized PVC has an impact strength of 3 kJ/m² at 20°C and 10 kJ/m² at 40°C. Is the material becoming more or less brittle as the temperature is increased?

Because there is an increase in the impact energy the material is becoming more ductile.

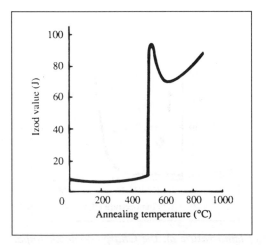

Figure 3.17   *Effect of annealing temperature on Izod test values*

3.5 Toughness test

Fracture toughness testing involves test pieces with sharp notches being strained until the crack propagates and the test piece fails. The problem in obtaining test pieces is producing the sharp notches. This is done by taking a test piece with a machined notch and then using a standardized pre-cracking procedure by loading with an alternating stress (fatigue loading) in order to obtain a sharp crack at the base of the machined notch.

Figure 3.18 shows forms of test pieces, as specified by BS 7448. The rectangular and square cross-section test pieces are for fracturing by three-point bending while the straight-notch compact and the stepped notch compact test pieces are for fracturing by tensile loading. For further details of test piece sizes, the production of the sharp notch and the calculation of the toughness from the test data, the reader is referred to BS 7448.

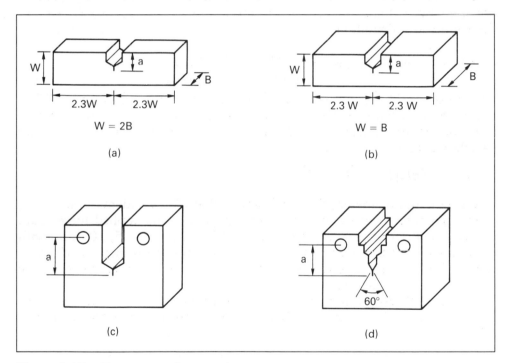

Figure 3.18   *Fracture toughness testing*

3.6 Hardness tests

The hardness of a material may be specified in terms of some standard test involving indentation or scratching of the surface of the material. The harder a material, the more difficult it is to make an indentation or a scratch. Hardness is essentially the resistance of the surface of a material to deformation. There is no absolute scale for hardness, each hardness form of test having its own scale. Though some relationships exist between results on one scale and those on another, care has to be taken in making comparisons because the different types of test are measuring different things.

The most common form of hardness tests for metals involves standard indenters being pressed into the surface of the material concerned. Measurements associated with the indentation are then taken as a measure of the hardness of the surface. The Brinell, Vickers and Rockwell tests are the main forms of such tests.

3.6.1 The Brinell test

With the Brinell test, a hardened steel ball is pressed for a time of 10 to 15 s into the surface of the material by a standard force (Figure 3.19). After the load and ball have been removed, the diameter of the indentation is measured. The Brinell hardness number (signified by HB) is obtained by dividing the size of the force applied by the surface area of the spherical indentation.

$$\text{Brinell hardness number} = \frac{\text{applied force}}{\text{area of indentation}}$$

The units used for the area are mm^2 and for the force kgf (1 kgf = 9.8 N and is the gravitational force exerted by 1 kg). The area can be obtained, from the measured diameter of the indentation and ball diameter, by either calculation or the use of tables.

$$\text{Area} = \tfrac{1}{2}\pi D\left[D - \sqrt{(D^2 - d^2)}\right]$$

where D is the diameter of the ball and d that of the indentation.

The diameter D of the ball used and the size of the applied force F are chosen, for the British Standard, to give F/D^2 values of 1, 5, 10 or 30 with the diameters of the balls being 1, 2, 5 or 10 mm. In principle, the same value of F/D^2 should give the same hardness value, regardless of the diameter of the ball used.

The Brinell test cannot be used with very soft or very hard materials. In the one case the indentation becomes equal to the diameter of the ball

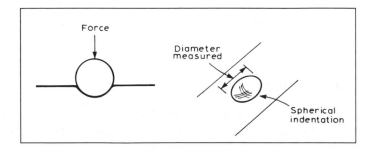

Figure 3.19   *The basis of the Brinell hardness test*

and in the other there is either no or little indentation on which measurements can be based. The thickness of the material being tested should be at least ten times the depth of the indentation if the results are not to be affected by the thickness of the material.

3.6.2 The Vickers test

The Vickers hardness test involves a diamond indenter being pressed under load for 10 to 15 s into the surface of the material under test (Figure 3.20). The result is a square-shaped impression. After the load and indenter are removed the diagonals d of the indentation are measured. The Vickers hardness number (HV) is obtained by dividing the size of the force, in units of kgf, applied by the surface area, in mm², of the indentation.

$$\text{Vickers hardness} = \frac{\text{applied force}}{\text{area of indentation}}$$

The surface area can be calculated, the indentation being assumed to be a right pyramid with a square base and an apex angle θ of 136°, or obtained by using tables and the diagonal values.

$$\text{Area} = \frac{d^2}{2 \sin \theta/2} = \frac{d^2}{1.854}$$

The Vickers test has the advantage over the Brinell test of the increased accuracy that is possible in determining the diagonals of a square as opposed to the diameter of a circle. Otherwise it has the same limitations as the Brinell test.

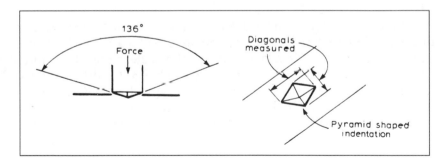

Figure 3.20   *The basis of the Vickers hardness test*

3.6.3 The Rockwell test

The Rockwell hardness test differs from the Brinell and Vickers tests in not obtaining a value for the hardness in terms of the area of an indentation but using the depth of indentation. The test uses either a diamond cone or a hardened steel ball as the indenter (Figure 3.21). A force of

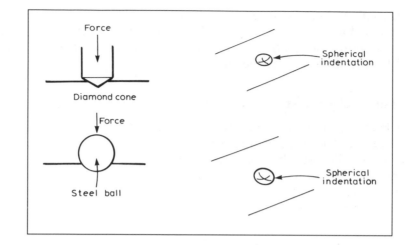

Figure 3.21   *The basis of the Rockwell hardness test*

90.8 N (10 kgf) is applied to press the indenter into contact with the surface. A further force is then applied and causes an increase in depth of the indenter's penetration into the material. The additional force is then removed and there is some reduction in the depth of the indenter due to the deformation of the material not being entirely plastic. The difference in the final depth of the indenter and the initial depth, before the additional force was applied, is determined. This is the permanent increase in penetration e due to the additional force. The Rockwell hardness number is then given by

Rockwell hardness number (HR) = $E - e$

where E is a constant determined by the form of the indenter. For the diamond cone indenter E is 100, for the steel ball 130.

There are a number of Rockwell scales, the scale being determined by the indenter and the additional force used. Table 3.2 indicates the scales and the types of materials for which each is typically used. In any reference to the results of a Rockwell test the scale letter must be quoted. For metals the B and C scales are probably the most commonly used ones.

For the most generally used indenters with the Rockwell test the size of the indentation is rather small. Localized variations of structure, composition and roughness can thus affect the results. The Rockwell test is, however, more suitable for workshop or 'on-site' use as it less affected by surface conditions than the Brinell or Vickers tests, which require flat and polished surfaces to permit accurate measurements.

A variation of the Rockwell test has to be used for thin sheet, this test being referred to as the *Rockwell superficial hardness test*. Smaller forces are used and the depth of indentation which is correspondingly smaller is measured with a more sensitive device. The initial force used is 29.4 N. Table 3.3 indicates the scales given by this test.

Table 3.2   *Rockwell hardness scales*

Scale	Indenter	Additional load (kg)	Typical applications
A	Diamond	60	Extremely hard materials, e.g. tool steels
B	Ball 1.588 mm dia.	100	Soft materials, e.g. Cu alloys, Al alloys
C	Diamond	150	Hard materials, e.g. steels, hard cast irons
D	Diamond	100	Medium case-hardened materials
E	Ball 3.175 mm dia.	100	Soft materials, e.g. Al alloys, Mg alloys, bearing metals
F	Ball 1.588 mm dia.	60	As E, the smaller ball being more appropriate where inhomogeneities exist
G	Ball 1.588 mm dia.	150	Malleable irons, gun metals, bronzes
H	Ball 3.175 mm dia.	60	Aluminium, lead, zinc
K	Ball 3.175 mm dia.	150	Al and Mg alloys
L	Ball 6.350 mm dia.	60	Plastics
M	Ball 6.350 mm dia.	100	Plastics
P	Ball 6.350 mm dia.	150	Plastics
R	Ball 12.70 mm dia.	60	Plastics
S	Ball 12.70 mm dia.	100	Plastics
V	Ball 12.70 mm dia.	150	Plastics

Note: the diameters of the balls arise from standard sizes in inches, 1.588 mm ¹⁄₁₆ in, 3.175 mm ⅛ in, 6.350 mm ¼ in, and 12.70 mm ½ in.

Table 3.3   *Rockwell superficial hardness scales*

Scale	Indenter	Additional load (kg)
15-N	Diamond	15
30-N	Diamond	30
45-N	Diamond	45
15-T	Ball 1.588 mm dia.	15
30-T	Ball 1.588 mm dia.	30
45-T	Ball 1.588 mm dia.	45

Note: the N scales are used for materials that, if thick enough, would have been tested on the C-scale, the T-scales for those on the B-scale.

3.6.4   Comparison of the different hardness scales

The Brinell and Vickers tests both involve measurements of the surface area of indentations, the forms of the indenters used being different. The Rockwell test entails measurements of the depth of penetration of indenters. Thus the various tests are concerned with different measurements as an indication of hardness. Consequently the values given by the different methods differ for the same material. There are no simple theoretical relationships between the various hardness scales, though some simple approximate, experimentally derived, relationships have been obtained. Different relationships, however, hold for different metals. The relationships are often presented in the form of tables. Table 3.4 shows part of a table for steels. Up to a hardness value of 300 the Vickers and Brinell values are almost identical.

There is an approximate relationship between hardness values and tensile strengths. Thus for annealed steels the tensile strength in MPa is about 3.54 times the Brinell hardness value, and for quenched and tempered steels 3.24 times the Brinell hardness value. For brass the factor is about 5.6 and for aluminium alloys about 4.2.

Example
An aluminium alloy has a hardness of 45 HB when annealed and 100 HB when solution treated and precipitation hardened. What might be the tensile strengths of the alloys in these two conditions if a factor of 4.2 is assumed?

Table 3.4 *Comparison of hardness scales for steels*

Brinell value	Vickers value	Rockwell B	Rockwell C
112	114	66	
121	121	70	
131	137	74	
140	148	78	
153	162	82	
166	175	86	4
174	182	88	7
183	192	90	9
192	202	92	12
202	213	94	14
210	222	96	17
228	240	98	20
248	248	102	24
262	263	103	26
285	287	105	30
302	305	107	32
321	327	108	34
341	350	109	36
370	392		40
390	412		42
410	435		44
431	459		46
452	485		48
475	510		50
500	545		52

Using a factor of 4.2, then the tensile strength in the annealed condition is 4.2 × 45 = 189 MPa. For the heat-treated condition it is 4.2 × 100 = 420 MPa. The measured values were 180 MPa and 430 MPa.

3.6.5 Hardness measurements with plastics

The Brinell, Vickers and Rockwell tests can be used with plastics. The Rockwell test with its measurement of penetration depth, rather than surface area, is more widely used. Scale R is a commonly used scale.

Another test that is used with plastics (BS 2782: Part 3) involves an indenter, a ball of diameter 2.38 mm, being pressed against the plastic by an initial force of 0.294 N for 5 s and then an additional force of 5.25 N being applied for 30 s. The difference between the two penetration depths is measured and expressed as a *softness number*. This is just the depth expressed in units of 0.01 mm. Thus a difference in penetration of 0.05 mm is a softness number of 5. The test is carried out at a temperature of 23 ± 1°C.

3.6.6 The Moh scale of hardness

A completely different form of hardness test, called the *Moh scale*, is based on assessing the resistance of a material to being scratched. Ten styli with points made of different materials are used. The styli materials are arranged in a scale so that each will scratch the one preceding it in the scale but not the one that follows it. The scale and materials are:

1 Talc
2 Gypsum
3 Calcspar
4 Fluorspar
5 Apatite
6 Felspar
7 Quartz
8 Topas
9 Corundum
10 Diamond

Thus, for example, felspar will scratch apatite but not quartz. Diamond will scratch all the materials while talc will scratch none of them. In this test the various styli are used until the lowest-number stylus is found that will just scratch it. The hardness number is then one less, since it is the number of the stylus that just fails to scratch the material. For example, glass can just be scratched by felspar but not by apatite. The glass thus has a hardness number of 5.

3.6.7 Hardness values

Figure 3.22 shows the general range of hardness values for different types of material when measured by the Vickers, Brinell, Rockwell and Moh test methods.

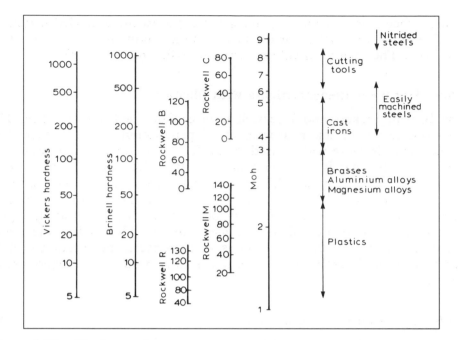

Figure 3.22 *Hardness values*

3.7 Electrical tests

The electrical resistivity or conductivity of a material requires a measurement of the resistance of a strip or block of the material. The British Standard for resistivity measurements with metals is BS 5714. In the case of metals the resistivity is very low and so the resistance to be measured can be low. For example, the resistance of a 1 m length of copper wire with a diameter of 1 mm is about 0.03 Ω at 20°C. Such a resistance is not easy to measure, since the means by which it is connected to the measurement system can have resistances of the same order of size or even larger. A smaller-gauge wire of 0.1 mm gives a resistance of about 2.1 Ω and is easier to measure. For routine measurements with resistances greater than 1 Ω the test piece can be what is termed a *two-terminal device*, i.e. there is just a single terminal at each end of the test piece for connections. For resistances of less than 1 Ω the test piece should be a *four-terminal device*, i.e. there are two terminals at each end. This means that the circuit connections to each end give less ambiguity as to between which points measurements are being made. For routine resistance measurements the method used should be capable of an accuracy of at least ± 0.30%. The method used for such resistance measurements is likely to be a resistance bridge, with possibly a Kelvin double bridge for small resistances.

In addition to measuring the resistance, the length and cross-sectional area of the test piece is required. The area can be obtained by direct measurement, but an alternative method which is often used is to weigh the test piece and calculate the area from a knowledge of the density and length, the area being mass/(density $\times$ length).

Since resistivity changes with temperature, it is important that the temperature t at which a measurement is made is noted. The following equation can then be used to correct the result to the reference temperature t_0 at which the result is required.

$$\rho_{t0} = \frac{\rho_t}{1 + (\alpha + \gamma)(t - t_0)}$$

where α is the temperature coefficient of resistance at the reference temperature and γ the coefficient of linear expansion.

For materials such as plastics or ceramics, the problem is that they have very high resistivities. This can present the problem that the surface layers, perhaps as a result of the absorption of moisture, might have a significantly lower resistivity than the bulk of the material and so the value indicated by the measurement is not that of the bulk material. Polymers also present the problem that when a voltage is applied across a sample the current through the material slowly decreases with time. Thus resistivity measurements need to have a time quoted with them, e.g. the value one minute after the application of a voltage.

3.7.1 Dielectric strength

The dielectric strength is the voltage needed per unit thickness of the material for electric breakdown (see Section 1.4). This can be measured by placing a sheet of the material between two conductors and increasing the voltage between them until there is an increase in the current from a barely measurable value to quite a significant current. The increase in current occurs because at breakdown the material changes from being a very good insulator to a quite reasonable conductor.

3.8 Chemical property tests

Materials in service can be affected by their environment and properties, such as a change in the mechanical properties of strength and toughness. Measurements of these properties thus give an indication of the interaction of the environment with a material.

Metallic materials corrode in moist air, with some metals corroding at a faster rate than others. Corrosion testing is not a simple operation with field trials or such conditions simulated in the laboratory. Essentially the tests are the observation of what happens to the metals over a period of time. Metals exposed to corrosive environments are often protected by being coated with a material such as paint. Tests are then used to investigate the weathering characteristics of the painted material. An accelerated weathering process is often used with exposure to radiation from an electric arc or with intermittent exposure to a spray of water to simulate rain. BS 6917 gives details of corrosion testing in artificial atmospheres, indicating the requirements for specimens, apparatus and procedures. BS 3900: Part G gives details of environmental tests on paint films.

The use of metals at high temperatures is often restricted by surface attack or scaling which gradually reduces the cross-sectional area and hence the stress-bearing ability of the item. The build-up of oxide layers at high temperatures is very much influenced by the environment, e.g. metal pipes exposed to superheated steam or hot gases from furnaces. Materials are tested by exposing them to such situations and measuring the reduction in the metal thickness as a consequence of the corrosive attack.

Plastic materials may dissolve in some liquids or absorb sufficient of the liquid to have their properties changed. When absorption occurs the plastic becomes permeable to the liquid, i.e. liquid can leak through it. This permeability is of vital concern if the plastic is being considered for used as a container for liquids, e.g. a Coca-Cola bottle (see Section 1.1).

Plastics are not generally subject to corrosion in the same way as metals but they can be adversely affected by weathering, i.e. exposure to light, heat or rain, sun. This can show itself as a fading of the colour of the plastic and/or a loss of flexibility. Tests, similar to those conducted on metals, are used to determine the weathering resistance of plastics. The tests tend to be comparative ones with standard colours/materials being simultaneously exposed and performances compared.

Example
Tables indicate that a weight loss of 1 mg per exposed area of 0.01 m² (1 dm²) per day for cast iron is a penetration of corrosion into the cast iron surface of 4.65 μm per year. What will be the penetration of corrosion into a cast iron product in a monitored situation if it suffers a weight loss of 0.5 mg/dm² in a day?

Since 1 mg/dm² in a day is 4.65 μm per year then 0.5 mg/dm² is 2.325 μm per year.

Example
The following data are test results on the corrosion rate for different metals suspended in the hot fumes from the combustion of fuel oils. On the basis of those data, coupled with the additional information supplied, discuss the possible choice of a metal for pipes which would be exposed to the fumes.

	Corrosion rate in mm/year
Steel with 25% Cr, 20% Ni	Completely corroded
35% Cr–65% Ni alloy	Completely corroded
50% Cr–50% Ni alloy	4
60% Cr–40% Ni alloy	2

The 50% Cr–50% Ni alloy has a tensile strength of 550 MPa, a yield stress of 340 MPa and a Charpy impact strength of 37 J. The 60% Cr–40% Ni alloy has a tensile strength of 760 MPa, a yield stress of 590 MPa and a Charpy impact strength of 7 J.

On the basis of the corrosion tests the choice is between the 50% Cr-65% Ni and the 60% Cr-40% Ni alloys, with the latter having better corrosion properties. The mechanical properties indicate that the 60% Cr-40% Ni alloy is stronger but considerably more brittle. The more ductile and shock-resistant properties are likely to mean that the 50% Cr-50% Ni alloy is the choice.

Problems

1 The following results were obtained from a tensile test of an aluminium alloy. The test piece had a diameter of 11.28 mm and a gauge length of 56 mm. Plot the stress–strain graph and determine (a) the tensile modulus and (b) the 0.1% proof stress.

Load/kN	0	2.5	5.0	7.5	10.0	12.5	15.0	17.5
Ext./mm	0	1.8	4.0	6.2	8.4	10.0	12.5	14.6

Load/kN	20.0	22.5	25.0	27.5	30.0	32.5	35.0
Ext./mm	16.3	19.0	21.2	23.5	25.7	28.1	31.5

Load/kN	37.5	38.5	39.0	39.0	(broke)
Ext./mm	35.0	40.0	61.0	86	

2 The following results were obtained from a tensile test of a polymer. The test piece had a width of 20 mm, a thickness of 3 mm and a gauge length of 80 mm. Plot the stress–strain graph and determine (a) the tensile strength and (b) the secant modulus at 0.2% strain.

Load/N	0	100	200	300	400	500	600	650	630
Ext./mm	0	0.08	0.17	0.35	0.59	0.88	1.33	2.00	2.40

3 The following results were obtained from a tensile test of a steel. The test piece had a diameter of 10 mm and a gauge length of 50 mm. Plot the stress–strain graph and determine (a) the tensile strength; (b) the yield stress and (c) the tensile modulus.

Load/kN	0	5	10	15	20	25	30	32.5
Ext./mm	0	0.016	0.033	0.049	0.065	0.081	0.097	0.106

Load/kN	35.8
Ext./mm	0.250

4 A flat tensile test piece of steel has a gauge length of 100.0 mm. After fracture, the gauge length was 131.1 mm. What is the percentage elongation?

5 The following data were obtained from a tensile test on a stainless steel test piece. Determine (a) the limit of proportionality stress; (b) the tensile modulus and (c) the 0.2% proof stress.

Stress/MPa	0	90	170	255	345	495	605
Strain/$\times 10^{-4}$	0	5	10	15	20	30	40

Stress/MPa	700	760	805	845	880	895
Strain/$\times 10^{-4}$	50	60	70	80	90	100

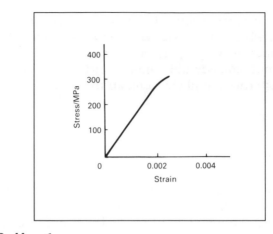

Figure 3.23 *Problem 6*

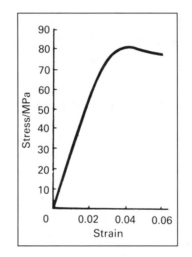

Figure 3.24 *Problem 7*

6   Estimate from the stress–strain graph for cast iron in Figure 3.23 the tensile strength and the limit of proportionality.

7   Estimate from the stress–strain graph for a sample of nylon 6 given in Figure 3.24 the tensile modulus and the tensile strength.

8   Sketch the form of the stress–strain graphs for (a) brittle stiff materials; (b) brittle non-stiff materials; (c) ductile stiff materials and (d) ductile non-stiff materials.

9   The effect of working an aluminium alloy (1.25% Mn) is to change the tensile strength from 110 MPa to 180 MPa and the elongation from 30% to 3%. What is the effect of the working on the properties of the material?

10   An annealed titanium alloy has a tensile strength of 880 MPa and an elongation of 16%. An annealed nickel alloy has a tensile strength of

700 MPa and an elongation of 35%. Which alloy is (a) the stronger and (b) the more ductile in the annealed condition?

11 Cellulose acetate has a tensile modulus of 1.5 GPa and polythene a modulus of 0.6 GPa. Which of the two plastics will be the stiffer?

12 The following are Izod impact energies at different temperatures for samples of annealed cartridge brass (70% Cu-30% Zn). What can be deduced from the results?

Temperature (°C)	+27	−78	−197
Impact energy (J)	88	92	108

13 The following are Charpy V-notch impact energies for annealed titanium at different temperatures. What can be deduced from the results?

Temperature (°C)	+27	−78	−196
Impact energy (J)	24	19	15

14 The following are Charpy impact strengths for nylon 6.6 at different temperatures. What can be deduced from the results?

Temperature (°C)	−23	−33	−43	−63
Impact strength (kJ/m²)	24	13	11	8

15 The impact strengths of samples of nylon 6, at a temperature of 22°C, are found to be 3 kJ/m² in the as-moulded condition but 25 kJ/m² when the sample has gained 2.5% in weight through water absorption. What can be deduced from the results?

16 With the Vickers hardness test a 30 kg load gave for a sample of steel an indentation with diagonals having mean lengths of 0.530 mm. What is the hardness?

17 With the Vickers hardness test a 30 kg load gave for a sample of steel an indention with diagonals having mean lengths of 0.450 mm. What is the hardness?

18 With the Vickers hardness test a 10 kg load gave for a sample of brass an indentation with diagonals having mean lengths of 0.510 mm. What is the hardness?

19 With the Brinell hardness test a 10 mm diameter ball and 3000 kg load gave an indentation with a diameter of 4.10 mm. What is the hardness?

20 With the Brinell hardness test a sample of cold-worked copper with a 1 mm diameter ball and 20 kg load gave an indentation of diameter 0.630 mm. What is the hardness?

21 The dielectric strength of a plastic was measured as 31 kV/mm when dry and after 2 days exposed to 80% humidity as 29 kV/mm. Explain the significance of the data.

22 After 4000 hours' exposure to the fumes in an oil-fired furnace samples of metals were found to show the following corrosion rate. Explain the significance of the data.

Steel 25%Cr-12%Ni 0.11 mm/year
Steel 25%Cr-20%Ni 0.28 mm/year
65%Ni-35%Cr alloy 0.02 mm/year

23   Oxide penetration on steels exposed for 5000 hours to steam at about 600°C was found to be as follows. Discuss the significance of the data.

0.11%C steel	0.40 mm
0.34%C steel	0.25 mm
1.24%Cr-0.5%Mo-1.4%Si steel	0.12 mm
2.25%Cr-0.5%Mo-0.75%Si steel	0.09 mm

24   The corrosion rate for mild steel test plates was found to give averages of 0.050 mm per year in rural surroundings, 0.070 mm per year in marine surroundings and 0.150 mm per year in heavy industrial surroundings. Discuss the significance of the data.

25   Specify the type of test that can be used in the following instances:

(a) A storekeeper has mixed up two batches of steel, one batch having been surface hardened and the other not. How could the two be distinguished?

(b) What test could be used to check whether tempering has been correctly carried out for a steel?

(c) A plastic is modified by the inclusion of glass fibres. What test can be used to determine whether this has made the plastic stiffer?

(d) What test could be used to determine whether a metal has been correctly heat treated?

(e) What test could be used to determine whether a metal is in a suitable condition for forming by bending?

<table>
<tr><td>

4

</td><td>

Structure and properties

</td></tr>
</table>

Outcomes At the end of this chapter you should be able to:

- Discuss the basic structure of metals, polymers and composites and the factors which affect the structure.
- Explain changes in properties in terms of changes in structure.
- Associate mechanical properties with particular structures.
- Identify structures which can lead to required properties.

4.1 Structure of metals

The term *metal* is used for elements, such as copper, which have atoms which so readily lose electrons that in the solid state at room temperature there are many free electrons. Thus in the solid state copper consists of an array of atoms each of which has lost one electron (Figure 4.1). This leaves each copper atom as having a net positive charge and it is termed a *positive ion*. The electrons that have been lost do not combine with any

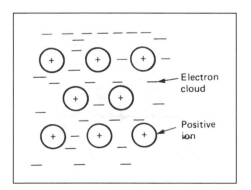

Figure 4.1 *Metals*

one ion but remain as a cloud of negative charge floating between the ions. The result is rather like a glue in that the cloud of electrons holds the positive ions together.

A simple model we can use to describe the structure of metals is to think of the ions in a metal being like spheres. Since the bonds formed between the positive ions can be formed in any direction without any restrictions, the only rule on how the spheres can be arranged is that imposed by how the geometry of the sphere restricts the packing together of spheres in order to give a tightly packed structure.

The free electrons explain why metals are good conductors of electricity, since they have free charge carriers which are easily moved through the solid by the application of a voltage. Insulators have no free electrons and the atoms in the solid are bonded together in a different way.

4.1.1   Crystals

One of the simplest arrangement of spheres is that of the *simple cubic structure*. Figure 4.2 shows the structure obtained by stacking four spheres with the centres of each sphere at the corners of a cube. The surfaces of each sphere touch the surfaces of each of its neighbours in such a way that the length of the side of the cube is equal to the diameter of the spheres. The dotted line in the Figure encloses what is termed the *unit cell*. To form a solid we can consider that this structure is repeated many times. The resulting solid would consist of a completely orderly array of spheres, i.e. atoms. We would expect the surfaces of such a solid to be smooth and flat with the angles between adjoining faces always 90°. Such a solid would, when broken up, always have the appearance of stacked cubes. This is a description of a cubic crystal.

A crystal thus consists of a large number of particles arranged in a regular repetitive array. It is this regularity which is characteristic of crystalline material. A solid having no such order in the arrangement of its constituent particles is said to be *amorphous*.

The simple cubic crystal shape is arrived at by stacking spheres in one particular way. It, however, is not the way that spheres can be most closely packed. By stacking spheres in a closer manner, as in Figure 4.3, other crystal shapes can be produced. With the *body-centred cubic* unit cell

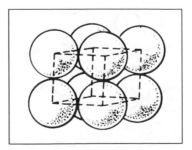

Figure 4.2   *A simple cubic structure*

(Figure 4.3(a)) the arrangement is slightly more complex than the simple cubic unit cell in having an extra sphere in the centre of the cell. With the *face-centred cubic* unit cell (Figure 4.3(c)) there is, when compared with the simple cubic unit cell, a sphere at the centre of each face of the cube. With the *hexagonal close-packed* unit cell (Figure 4.3(b)) the spheres are packed in a close array which gives a hexagonal form of structure. These are just the three closest packed arrangements which we can make from packing identical spheres together. Because metals can be considered to be composed of spherical atoms, it is these three close-packed structures which are used for solid metals.

An important point to note with all these structures is that there are spaces between the spheres in the crystal structures. The size of these

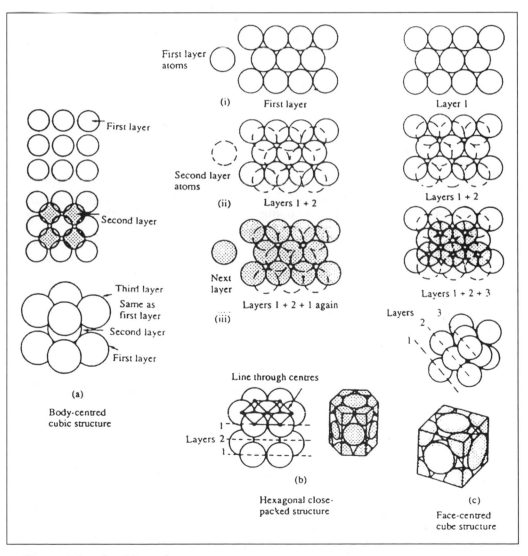

Figure 4.3 *Stacking spheres*

57

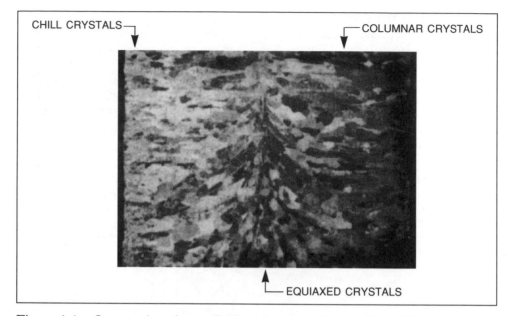

Figure 4.4 *Cross-section of a small aluminium ingot (adapted from H.A. Monks and D.C. Rochester,* Technician Structure and Properties of Metals, *Cassell)*

spaces depends on the type of structure. Within these spaces it is possible to fit other atoms, provided they are small enough, without too much strain on the crystalline structure. In some cases, with some strain, atoms can be forced into spaces which are really too small for them. This is discussed later in this chapter in connection with alloys.

4.1.2 Metals as crystalline

Metals are crystalline substances. This may seem a strange statement in that metals do not generally look like crystals, with their geometrically regular shapes. However, if we consider a metal in solidifying from the liquid as not growing as a single crystal but having crystals starting to grow at a large number of points within the liquid then the result is a mass of crystals. Each crystal, in growing, is then prevented from reaching geometrically regular shapes by neighbouring crystals restricting its growth. Figure 4.4 shows a section of a metal and reveals such a mass of crystals.

The term *grain* is used to describe the crystals within the metal. A grain is merely a crystal without its geometrical shape and flat surfaces because its growth was impeded by contact with other crystals. Within a grain the arrangement of particles is just as regular and repetitive as within a crystal with smooth faces. A simple model of a metal with grains is given if a raft of bubbles is produced on the surface of a liquid (Figure 4.5). The bubbles pack together in an orderly and repetitive manner but if 'growth' is started at a number of centres then 'grains' are produced. At the boundaries between the 'grains' the regular pattern breaks down as the pattern changes from the orderly one of one 'grain' to that of the next 'grain'.

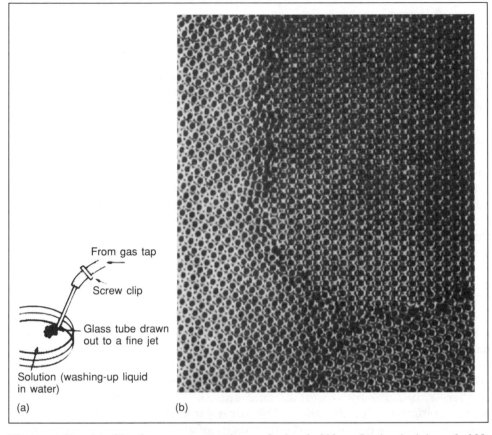

From gas tap

Screw clip

Glass tube drawn
out to a fine jet

Solution (washing-up liquid
in water)

(a)   (b)

Figure 4.5   *(a) Simple arrangement for producing bubbles; (b) 'grains' in a bubble raft (adapted from the Royal Society)*

The grains in the surface of a metal are not generally visible, though an exception is the very large grains which are readily visible in the surface of galvanized steel objects. Grains can, however, be made visible by careful etching of the polished surface of the metal with a suitable chemical. The chemical preferentially attacks the grain boundaries. For example, in the case of copper and its alloys, concentrated nitric acid can be used. In the case of carbon and alloy steels of medium carbon content a etchant called nital can be used. Nital is a mixture of nitric acid and alcohol, typically 5 ml of acid to 95 ml of alcohol. Details of suitable chemicals are given in reference books, e.g. *Metals Databook* by C. Robb (The Institute of Metals), *ASM Metals Reference Book* (American Society for Metals). However, proper safety precautions in the handling and disposing of the chemicals are vital since they are highly corrosive and many of them are potentially lethal (see Chapter 7).

Examples of metals which, in the pure state, adopt the body-centred cubic unit cell form of structure are iron, chromium and molybdenum, face-centred cubic unit cell forms of structure are aluminium, copper, lead and nickel, with the hexagonal close-packed unit cell being given by magnesium and zinc.

4.2 Alloys

An *alloy* is a metallic material made by mixing two or more elements. The everyday metallic objects around you will be made, almost invariably, from alloys rather than the pure metals themselves. Pure metals do not always have the appropriate combinations of properties needed; alloys can, however, be designed to have them.

Making alloys is rather like baking a cake. The basic ingredients of flour, sugar, fat, eggs and water are mixed together and then cooked. The result is a cake which has a texture and properties quite different from those of the individual ingredients. The type of cake produced depends on the relative amounts of the ingredients and the way it is cooked. In making alloys, the ingredients are mixed and heated and the resulting alloy can have properties quite different from those of the ingredients. The properties will depend on the relative amounts and nature of the ingredients as well as how they are 'baked'. An alloy is a particular mixture of components and so has a particular chemical composition, e.g. one carbon steel may be 99.0% iron combined with 1.0% carbon while another is 99.5% iron with 0.5% carbon.

The coins in your pocket are made of alloys. Coins need to be made of a relatively hard material which does not wear away rapidly, i.e. they must have a 'life' of many years. Coins made of, say, pure copper would be very soft; not only would they suffer considerable wear but they would bend in your pocket. The copper-looking British coins are made of an alloy of copper with 2.5% by weight of zinc and 0.5% of tin, the term *coinage bronze* being used for the alloy. The silver-looking British coins are made of an alloy of copper with 25% by weight of nickel, the term *cupro-nickel* being used.

Pure metals tend to be soft with high ductility, low tensile strength and low yield strength. Because of this they are rarely used in engineering. Alloying can produce harder materials with higher tensile strength, higher yield stress and a reduction in ductility. Such materials are more useful in engineering. There are, however, some circumstances in which the properties of pure metals are useful. These are where high electrical conductivity is required (alloying reduces conductivity); where good corrosion resistance is required (alloying can result in less corrosion resistance); and where very high ductility is required.

We can think of the structure of alloys in terms of the constituent metals, say A and B, being mixed in the liquid state. Then when the mixture solidifies, there is the possibility that solid alloy will have a crystal structure in which some of the atoms in the crystal structure of A have been replaced by atoms of B (Figure 4.6(a)). Alternatively, because there are spaces between the atoms of A in its crystal structure, some atoms of A, if small enough, might lodge in these spaces (Figure 4.6(b)). Another possibility is that elements A and B combine to form a chemical compound. With a compound there will be a particular structure for that compound with atoms of A and B assuming specific positions, rather than just popping into any gap. Another possibility is that when the liquid mixture cools A and B separate out, with B forming its own crystal

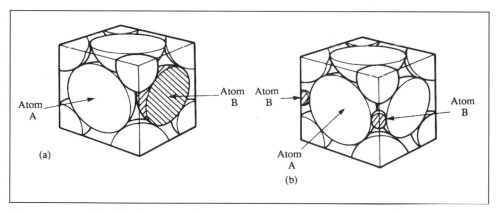

Figure 4.6 *Possible forms of alloys*

structure independent of A. The structure then becomes a mixture of two types of crystals.

4.2.1 Ferrous alloys

Pure iron is a relatively soft material and is hardly ever used. Alloys of iron with carbon are, however, very widely used, the term *ferrous alloys* being given to alloys with iron. Pure iron at room temperature exists as a body-centred cubic structure, this being commonly referred to as *ferrite*. This form continues to exist up to 912°C. At this temperature the structure changes to a face-centred cubic one, known as *austenite*. Iron atoms have a diameter of 0.256 nm (1 nanometre = 10^{-9} m), carbon atoms are much smaller with a diameter of 0.154 nm. The face-centred cubic structure is a more open structure than the body-centred cubic. The face-centred structure of austenite has voids which can accommodate spheres of up to 0.104 nm in diameter, the body-centred cubic structure has voids between the atoms which are 0.070 nm in diameter. Thus, carbon atoms can be more easily accommodated within austenite, without severe distortion of the lattice, than ferrite. Austenite can take up to 2.0% of carbon while ferrite can take only 0.2%. Thus when iron containing carbon is cooled from the austenite state to the ferrite state, there is a reduction in the amount of carbon that can be accommodated within the iron and so some of the carbon atoms come out of the crystals and form a compound, another crystal structure, between iron and carbon called *cementite*. Cementite is hard and brittle. The result can be a structure consisting of purely ferrite grains mixed with grains which have a laminated structure of ferrite and cementite. Such a laminated structure is termed *pearlite*. Pure cementite is harder than pearlite, which in turn is harder than pure ferrite. Thus the structure, and hence the properties, of the iron alloy is determined by the amount of carbon present.

The percentage of carbon alloyed with iron has a profound effect on the properties of the alloy. The terms used for the alloys produced with different percentages of carbon are:

Wrought iron	0 to 0.05% carbon
Steel	0.05 to 2% carbon
Cast iron	2 to 4.5% carbon

The term *carbon steel* is used for those steels in which essentially just iron and carbon are present. *Alloy steel* is used where other elements are included.

4.2.2 Plain carbon steels

Carbon steels are grouped according to their carbon content with the designations being roughly:

Mild steel	0.10 to 0.25% carbon
Medium-carbon steel	0.20% to 0.50% carbon
High-carbon steel	More than 0.50% carbon

Mild steel has a structure consisting predominantly of ferrite, medium-carbon steels tend to have about equal amounts of ferrite and pearlite, while high-carbon steels have predominantly pearlite with some free cementite occurring at high carbon contents.

Figure 4.7 shows how the mechanical properties of carbon steels depend on the percentage of carbon. Increasing the percentage of carbon, within the range considered, increases the amount of pearlite at the expense of the softer ferrite, and hence:

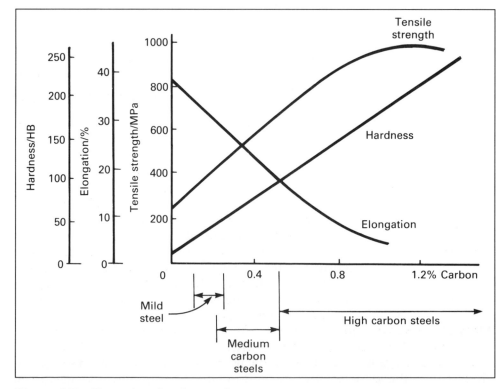

Figure 4.7 *Properties of carbon steels*

1 Increases the tensile strength
2 Increases the hardness
3 Decreases the percentage elongation
4 Decreases the impact strength

Mild steel is a general-purpose steel and is used where hardness and tensile strength are not the most important requirements but ductility is often needed. Typical applications are sections for use as joists in buildings, bodywork for cars and ships and screws, nails and wire. Medium-carbon steel is used for agricultural tools, fasteners, dynamo and motor shafts, crankshafts, connecting rods and gears. With such steels the lower ductility puts a limit on the types of processes that can be used. Medium-carbon steels are capable of being quenched and tempered to develop reasonable toughness with strength (see Section 5.3). High-carbon steel is used for withstanding wear, where hardness is a more necessary requirement than ductility. It is used for machine tools, saws, hammers, cold chisels, punches, axes, dies, taps, drills and razors. The main use of high-carbon steel is mainly as a tool steel. High-carbon steels are usually quenched and tempered at about 250°C to develop their high strength with some slight ductility (see Section 5.3).

Example
A pickaxe head may be made of a high-carbon steel. Why high-carbon rather than mild steel?

High-carbon steel is a harder, stronger material than mild steel. The higher ductility of mild steel is not required in this situation.

4.2.3 Non-ferrous alloys

The term *non-ferrous alloy* is used for all alloys where iron is not the main constituent, e.g. alloys of aluminium, of copper, of magnesium, etc. The following are some of the general properties and uses of non-ferrous alloys in common use in engineering.

Aluminium alloys	Low density, good electrical and thermal conductivity, high corrosion resistance. Tensile strengths of the order of 150 to 400 MPa, tensile modulus about 70 GPa. Used for metal boxes, cooking utensils, aircraft bodywork and parts.
Copper alloys	Good electrical and thermal conductivity, high corrosion resistance. Tensile strengths of the order of 180 to 300 MPa, tensile modulus about 20 to 28 GPa. Used for pump and valve parts, coins, instrument parts, springs, screws.
Magnesium alloys	Low density, good electrical and thermal conductivity. Tensile strengths of the order of 250 MPa and tensile modulus about 40 GPa. Used as castings and forgings in the aircraft industry where weight is an important consideration.

Nickel alloys	Good electrical and thermal conductivity, high corrosion resistance, can be used at high temperatures. Tensile strengths between about 350 and 1400 MPa, tensile modulus about 220 GPa. Used for pipes and containers in the chemical industry where high resistance to corrosive atmospheres is required, food processing equipment, gas turbine parts.
Titanium alloys	Low density, high strength, high corrosion resistance, can be used at high temperatures. Tensile strengths of the order of 1000 MPa, tensile modulus about 110 GPa. Used in aircraft for compressor discs, blades and casings, in chemical plant where high resistance to corrosive atmospheres is required.
Zinc alloys	Low melting points, good electrical and thermal conductivities, high corrosion resistance. Tensile strengths about 300 MPa, tensile modulus about 100 GPa. Used for car door handles, toys, car carburettor bodies – components that in general are produced by pouring the liquid metal into dies.

As an example of a non-ferrous alloy, consider copper alloys. Pure copper is a soft material with low tensile strength. For many engineering purposes it is alloyed with other metals. The exception is where high electrical conductivity is required. Pure copper has a better conductivity than the alloys. The following indicate the names given to the various types of copper alloys:

Copper with zinc	Brasses
Copper with tin	Bronzes
Copper with tin and phosphorus	Phosphor bronzes
Copper with tin and zinc	Gun metals
Copper with aluminium	Aluminium bronzes
Copper with nickel	Cupro-nickels
Copper with zinc and nickel	Nickel silvers
Copper and silicon	Silicon bronze
Copper and beryllium	Beryllium bronze

Figure 4.8 shows how the percentage of zinc included with brasses affects the mechanical properties. Brasses with between 5% and 20% zinc are called gilding metals and, as the name implies, are used for architectural and decorative items to give a 'gilded' or golden colour. Cartridge brass is copper with 30% zinc. One of its main uses is for cartridge cases, items which require high ductility for the deep drawing process used to make them. The term *basis brass* is use for copper with 37% zinc. This is a good alloy for general used with cold working processes and is used for fasteners and electrical connectors. It does not have the high ductility of those brasses with less zinc. Copper with 40% zinc is called Muntz metal.

The changes in the properties of brasses when the amount of zinc is altered arise from changes in the structure. Brasses with between 0% and

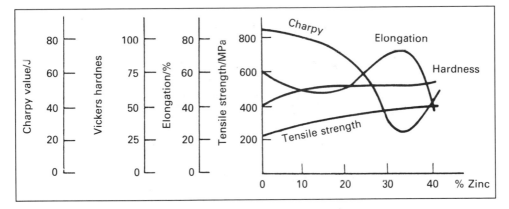

Figure 4.8   *Properties of brasses*

35% zinc form one type of structure, termed *alpha*, and between 5% and 45% there is a mixture of this alpha structure and another structure termed beta. It is this change in structure, i.e. the way the atoms of copper and zinc are packed together, that is responsible for the abrupt changes in properties of brass at 35% zinc.

4.3   Stretching metals

A simple way we can think of the atoms in a metal is as though they were an array of spheres tethered to each other by springs, as illustrated in Figure 4.9(a). When forces are applied to stretch the material then the springs are stretched and exert an attractive force pulling the atoms back to their original positions. When forces are applied to compress the material then the springs are compressed and exert repulsive forces which push the atoms back to their original positions (Figure 4.9(b)). We thus have a model for interatomic forces.

A simple theory to explain the elastic and plastic behaviour of metals when stretched is the *block slip theory*. Consider a block of atoms in the form suggested by the above model. In the absence of any externally applied forces the atoms are all in their equilibrium positions (Figure 4.10(a)). When stress is applied to a metal then we will consider that the block of atoms is at such an angle to the forces that the situation is as shown in Figure 4.10(b). Elastic strain occurs when the atoms all become displaced from their equilibrium positions and then spring back to them when the stress is removed. However, if the stress is high enough then yielding occurs and blocks of atoms slip (Figure 4.10(c)). When the stress is removed the atoms spring back to equilibrium positions but for some atoms these are new positions and permanent deformation has been produced (Figure 4.10(d)). The plane along which atoms slip is called the *slip plane*.

In terms of our model of a crystal as a pile of stacked spheres, we can consider that slip is when an entire row of spheres is pushed sufficiently

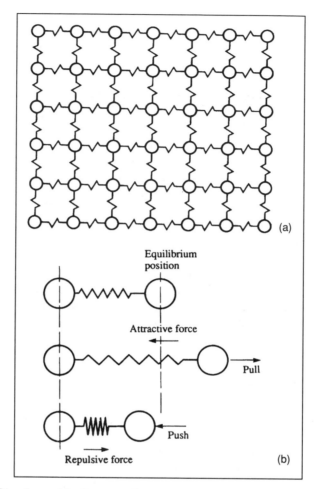

Figure 4.9 *Simple model of a crystal*

to all move along one position, as illustrated in Figure 4.11. On this model of slip, we can only have slip within an orderly arrangement of spheres, i.e. within a grain. Slip planes cannot thus cross over from one grain to another, the disorderly arrangement at the grain boundary does not allow it. Slip will only occur in those grains which have atomic planes at suitable angles to the applied forces. Thus a metal having big grains can result in more slippage than one having a larger number of small grains. A simple model we might use is to consider soldiers on parade in orderly ranks, i.e. all the soldiers in one large 'grain'. For the movement of one soldier in the back rank to step forward then all the soldiers in that line step forward and there is 'slip' and a large amount of movement. The analogy with the small grain metal is of a football crowd. When one person moves then there might be some local 'slip' as other people move but there is no overall movement of the crowd. Thus, with metals, the bigger the grains, the greater the amount of plastic deformation that might be expected. A fine-grain structure will have less

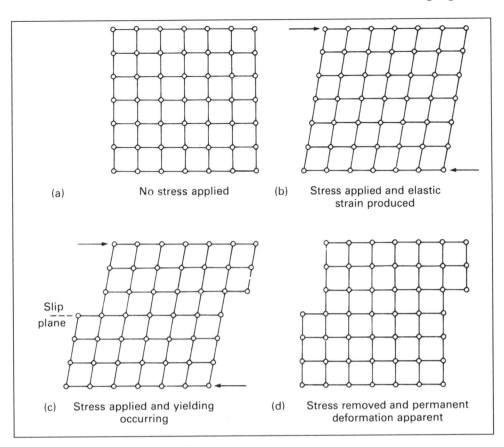

(a)   No stress applied   (b)   Stress applied and elastic strain produced

Slip plane

(c)   Stress applied and yielding occurring   (d)   Stress removed and permanent deformation apparent

Figure 4.10   *Block slip model*

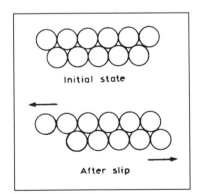

Initial state

After slip

Figure 4.11   *Slip of atom planes*

slippage and so show less plastic deformation, i.e. be less ductile. A brittle material is thus one in which each slip process is confined to a short run in the metal and not allowed to spread. A ductile material is one in which the slip process is not confined to a short run in the metal and does spread over a large part of it.

While there can be considered to be many planes of atoms in a crystal, slip is found to occur only between the planes with the closest packing of atoms. This is because the atoms are close enough to more easily permit changes of positions than when further apart. Figure 4.12 illustrates this concept of close-packed planes, the lines indicating the planes with the highest density of atoms per unit length. If you draw other lines through atoms you will find less atoms per unit length. The number of such high-density planes along which slip can occur depends on the form of structure of the crystal. The body-centred cubic structure has many such slip planes, the face-centred cubic less and the hexagonal close-packed structure even less. Thus metals which have a hexagonal close-packed structure tend to be harder and more brittle than those with the face-centred cubic structure, while the body-centred cubic structure is likely to be the least hard and most ductile metal.

The above is just a simple model of what happens when metals are stretched. The model needs modification to do more than give a simple idea of what happens. The above model has assumed that the arrangement of atoms within a grain is perfectly orderly. In reality this is not the case and there are some atoms in the wrong places, the term *dislocations* being used. Thus, within a grain, we might have the situation shown in Figure 4.13(a). When stress is applied the dislocation moves through the array of atoms, as illustrated in Figure 4.13(b)–(d) so the slip takes place

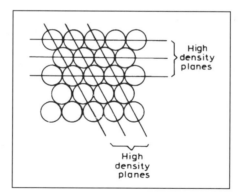

Figure 4.12   *High-density planes*

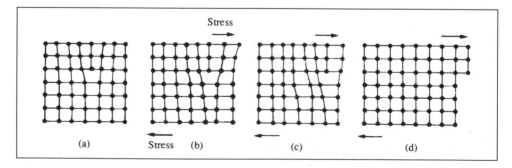

Figure 4.13   *Movement of a dislocation under the action of stress*

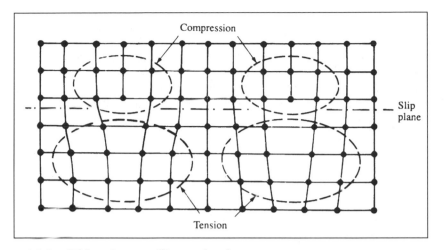

Figure 4.14 *Dislocations repelling each other*

atom by atom rather than the wholesale movement of one plane of atoms past another.

When the movement of a dislocation through a metal brings it to another dislocation then they can either cancel each other out or hinder further movement. Figure 4.14 shows what can happen when two dislocations come close to each other. Each dislocation has the atoms on one side of the slip plane in compression and on the other side in tension. When two compression regions come close together the forces between the atoms result in the dislocations repelling each other. In general, the more dislocations a metal has, the more the dislocations get in the way of each other and so the more difficult it is for the dislocations to move through the metal and hence slip to occur. Thus the greater the number of dislocations, the greater the stress needed to produce yielding.

Dislocations are produced as a result of missing atoms, atoms being displaced from their correct positions, and foreign atoms being present and distorting the orderly packing of atoms. Cold working, which distorts grains, results in an increase in dislocations because it displaces atoms from their correct positions. The foreign atoms may be present as a result of a deliberate alloying process. Thus alloying, in increasing the number of dislocations and making it more difficult for dislocations to move through the material, increases the yield stress. Another method of increasing the yield stress is to cause small particles to precipitate out from an alloy. Such a process is called *precipitation hardening*.

The above is just an indication of how the properties of metals can be explained in terms of their structure.

4.4 Cold working

Suppose you were to take a carbon steel test piece and perform a tensile test on it. You may, for instance, find that the material showed a yield

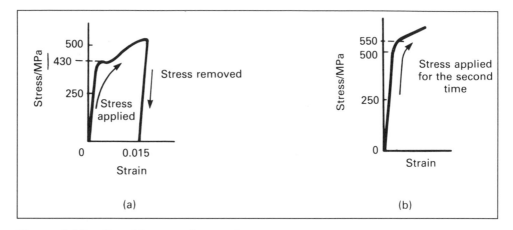

Figure 4.15 *Stretching a carbon steel*

stress of 430 MPa. If the stress is continued beyond this point but the stress released before the material breaks then a permanent deformation will be found to have occurred. Figure 4.15(a) illustrates this sequence of events and indicates a permanent deformation of strain of 0.015. Now suppose the material is again stretched. This time the yield stress is not 430 MPa but 550 MPa. The material has now a much higher yield stress (Figure 4.15(b)). It is not only the yield stress which has increased, the tensile strength has increased, the percentage elongation has decreased, and the hardness has increased.

The material is said to have been subject to *cold working* and the above are typical of the changes that occur. The term cold working is used when plastic deformation is produced at a temperature which is not high enough to produce changes. The term *work hardening* is often used since the cold working has resulted in the material becoming harder. Figure 4.16 shows the effect of cold working on the hardness of typical materials. The more a material is worked, the harder it becomes. A stage is reached, however, when the hardness is at a maximum and further deformation is not possible as the material is too brittle. Thus, for example, with the aluminium referred to in the Figure this is when the aluminium has been reduced in thickness by about 60%. The material is then said to be *fully work hardened*.

An example of cold working is the cold rolling of sheet to produce thinner sheet (Figure 4.17). With annealed aluminium sheet being rolled, full work hardening occurs with a reduction in thickness of about 60% and so is about the maximum that can be produced. The annealed sheet might typically have a tensile strength of 90 MPa, an elongation of 40% and a hardness of 20 HV before work hardening. Full work hardening might result in a tensile strength of 150 MPa, an elongation of 3% and a hardness of 40 HV.

We can offer an explanation for these effects of cold working on the properties of metals by considering that the plastic deformation results in slip. Thus, with rolling, the grains become elongated and distorted in the

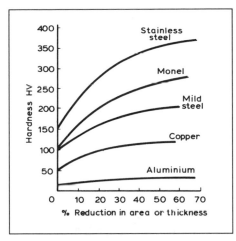

Figure 4.16   *The effect of cold working on hardness*

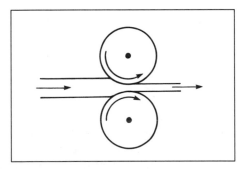

Figure 4.17   *Cold rolling*

direction of the rolling. There is therefore a difference in mechanical properties in the rolled sheet in the direction of rolling and at right angles to it. In addition, the number of dislocations of atoms within the grains increases. The result is a less orderly structure, so making it more difficult to produce further slip, hence the increase in brittleness.

Example
Using Figure 4.16, what is the approximate percentage reduction in thickness of a sheet of mild steel that is possible before it becomes fully work hardened?

The mild steel would appear to have reached its maximum hardness with a reduction in thickness of about 50–60% and so be fully work hardened.

4.4.1   Heat treating cold-worked metals

Cold working of metals results in changes in mechanical properties with yield stress, tensile strength, and hardness increasing and percentage

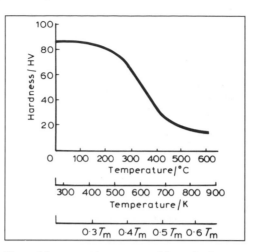

Figure 4.18   *The effect of heat treatment on cold-worked copper*

elongation decreasing. These changes can be reversed by suitable heat treatment.

Cold-worked metals generally have deformed grains, with a high density of dislocations within the grains. Such dislocations lead to internal stresses within grains. When such a metal is heated then there is some slight rearrangement of atoms within the grains and a consequent reduction in internal stresses. This process is known as *recovery*.

When a cold-worked metal is heated above about $0.3T_m$, where T_m is the melting point of the material on the kelvin scale of temperature, then there is a marked reduction in tensile strength, hardness and an increase in percentage elongation. Figure 4.18 illustrates this for cold-worked copper. What is happening is that the material is beginning to *recrystallize*. New grains start to grow. The temperature at which recrystallization starts is called the *recrystallization temperature*. For pure metals it tends to be about 0.3 to $0.5T_m$. Thus, aluminium which has a melting point of 933 K has a recrystallization temperature of 423 K, about $0.45T_m$. Iron with a melting point of 1356 K has a recrystallization temperature of 473 K, about $0.35T_m$. As the temperature is further increased so the crystals start to grow until they have completely replaced the original distorted cold worked structure. Figure 4.19 illustrates the sequence of events.

The term *annealing* is used for the heat treatment process which involves heating the material to above the recrystallization temperature and obtaining more ductile properties.

Example

A manufacturer of copper sheet receives the copper as much thicker plate, with already some amount of cold working. He proposes to produce the sheet by cold rolling in a number of stages, the stages being separated by annealing. Why is the sheet production in a number of stages?

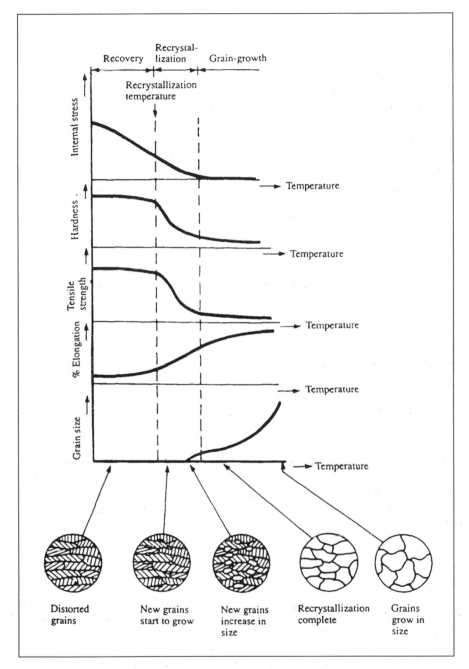

Figure 4.19   *The effect of heat treatment on cold-worked metals*

If copper is cold worked to about a 60% reduction in thickness it becomes brittle and tends to break with further working. Also it becomes fairly hard and difficult to roll. By following the rolling by annealing the material is made more ductile again and rolling can continue without difficulty.

4.5   Hot working

Cold working involves plastically deforming materials at temperatures which are below the recrystallization temperature. The result is a harder, less ductile material with deformed grains. *Hot working* involves deforming a material at a temperature greater than the recrystallization temperature. Then, as soon as a grain becomes deformed it recrystallizes. No hardening thus occurs and the working can be continued without any difficulty. No interruption of working is needed to anneal the material, as is the case with cold working.

A disadvantage of hot working is that oxidization of the metal surfaces occurs. Cold working does not have this problem. Another disadvantage is that the material will have comparatively low values of hardness and tensile strength, with high percentage elongation. A combination of hot and cold working is thus often used in a particular shaping process. The first operation, involving large amounts of plastic deformation, is carried out by hot working. After cleaning the surfaces of the metal, it is then cold worked to increase the strength and hardness and give a good surface finish.

Example

Lead has a melting point of 327°C. Will the product be work hardened if it is made by extruding at room temperature? Extrusion is a process rather similar to the squeezing of toothpaste out of a tube, the metal being squeezed out through a nozzle and taking the shape dictated by that of the nozzle.

The melting point of lead is about 600 K. This would mean that the extrusion at about 300 K is at about $0.5T_m$. In other words, the extrusion is taking place at about the recrystallization temperature. The process is likely to be just about a hot working process and so there would be no work hardening.

4.6   The structure of polymers

The plastic washing-up bowl, the plastic measuring rule, the plastic cup – these are all examples of materials which have polymer molecules as their basis. A polymer molecule in a plastic may have thousands of atoms all joined together in a long chain. The backbones of these long molecules are chains of carbon atoms. Carbon atoms are able to form strong bonds with themselves and produce long chains to which other atoms can become attached.

The term *polymer* is used to indicate that a compound consists of many repeated structural units. The prefix 'poly' means many. Each structural unit in the compound is called a *monomer*. For many plastics the monomer can be deduced by deleting the prefix 'poly' from its name. Thus the plastic called polyethylene is a polymer which has as its monomer ethylene. Figure 4.20 shows the monomer and the resulting polymer.

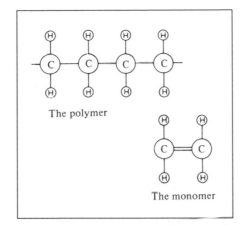

Figure 4.20   *Polyethylene*

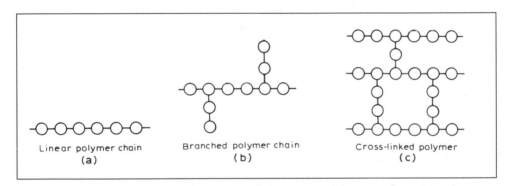

Figure 4.21   *(a) Linear polymer chain; (b) branched polymer chain; (c) cross-linked polymer chains*

Figure 4.21 shows the basic forms that can be adopted by the molecular chains. These forms can be described as linear, branched and cross-linked chains. The linear chains have no side branches or links with other chains and can thus move readily past each other. If, however, the chains have side branches there is a reduction in the ease with which chains can move past each other and so the material is more rigid. If there are cross-links a much more rigid material is produced in that the chains cannot slide past each other at all.

Polymers can be classified as thermoplastics, thermosets or elastomers (see Section 1.8.2). A simple method by which thermoplastics and thermosets can be distinguished is when heat is applied. With a thermoplastic the material softens with removal of the heat resulting in hardening. With a thermoset, heat causes the material to char and decompose with no softening. An elastomer is a polymer which by its structure allows considerable extensions which are reversible. Thermoplastics have linear or branched chains for their structure. Thermosets have a cross-linked structure. Elastomers are chains with some degree of cross-linking.

4.6.1 Additives

The term *plastic* is commonly used to describe materials based on polymers. Such materials, however, invariably contain other substances which are added to the polymers to give the required properties. Since some polymers are damaged by ultraviolet radiation, protracted exposure to the sun can lead to a deterioration of mechanical properties. An ultraviolet absorber is thus often added to the polymer, such an additive being called a *stabilizer*. Carbon black is often used for this purpose. *Plasticizers* are added to the polymer to make it more flexible. In one form this may be liquid which is dispersed throughout the solid, filling the space between the polymer chains and acting like a lubricant and permitting the chains to slide past each other more easily. *Flame retardants* may be added to improve fire-resistant properties and pigments and dyes to give colour to the material. The properties and cost of a plastic can be markedly affected by the addition of substances termed *fillers*. Since fillers are generally cheaper than the polymer, the overall cost of the plastic is reduced. Often up to 80% of a plastic may be filler. Examples of fillers are glass fibres to increase the tensile strength and impact strength, mica to improve electrical resistance, graphite to reduce friction and wood flour to increase tensile strength. One form of additive used is a gas to give foamed or expanded plastics. Expanded polystyrene is used as a lightweight packaging material and foamed polyurethane as a filling for upholstery and sponges.

4.7 Thermoplastics

Thermoplastics consist of polymers with long-chain molecules which are either linear chains or long chains with small branches. Linear chains have no side branches or cross-links with other chains. Because of this they can easily move past each other. If, however, the chain has branches then there is a reduction in the ease with which chains can be made to move past each other. This shows itself in the material being more rigid, i.e. less strain produced for a given stress.

A crystalline structure is one in which there is an orderly arrangement of particles and a structure in which the arrangement is completely random is said to be amorphous. Many polymers are amorphous with the polymer chains being completely randomly arranged in the material (Figure 4.22). Linear polymer molecules can, however, assume an arrangement which is orderly. Figure 4.23 shows the type of arrangement of chains that can occur, the linear chains folding backwards and forwards on themselves. The arrangement is said to be *crystalline*. The tendency of a polymer to crystallize is determined by the form of the polymer chains. Linear polymers can crystallize to quite an extent, but complete crystallization is not obtained in that there are invariably some regions of disorder. Polymers with side branches show less tendency to crystallize since the branches get in the way of the orderly arrangement. The greater the crystallinity of a polymer, the closer the polymer chains can be packed and so the greater the density.

Figure 4.22 *A linear amorphous polymer. Individual atoms are not shown, the chains being represented by lines*

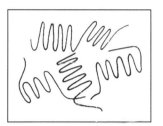

Figure 4.23 *Folded linear polymer chains*

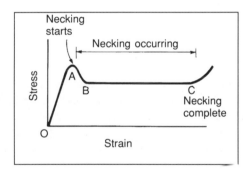

Figure 4.24 *Typical stress strain graph for a crystalline polymer*

Figure 4.24 shows the typical form of a stress–strain graph for a crystalline polymer. When stress is applied the first thing that begins to happen is that some movement of folded chains past each other occurs. However, when point A is reached the polymer chains start to unfold to give a material with the chains all lying along the direction of the forces stretching the material. The material shows this by starting to exhibit *necking* (Figure 4.25). As the stress is further increased, the necking spreads along the material with more and more chains unfolding. Eventually, when the entire material is at the necked stage all the chains have lined up. The material in this state behaves differently from earlier in the stress–strain graph, the material being said to be *cold drawn*. The above sequence of events tends to occur only if the material is stretched

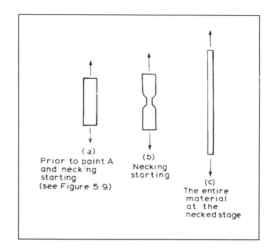

Figure 4.25 *Necking in a polymer*

slowly and sufficient time elapses for the molecular chains to unfold. If a high strain rate is used the material is likely to break without the chains all becoming lined up. The plastic used for making polythene bags is a crystalline polymer. Try cutting a strip of polythene from such a bag and pulling it between your hands and see the necking develop with low rates of strain.

4.7.1 Examples of thermoplastics

Examples of thermoplastics are polyethylene (polythene), polypropylene, polyvinyl chloride (PVC), polystyrene, acrylonitrile–butadiene–styrene terpolymer (ABS) and polyamides (nylons). The following is a brief discussion of the structure of these thermoplastics and their properties.

Polyethylene can exist in two forms, one as a linear chain and one as a chain with side branches. Figure 4.26 shows the two forms. With the linear chains a high degree of crystallization occurs and some 95% of the material will thus be packed in an orderly manner. This results in the material having a higher density than the branched form of polyethylene for which a crystallinity of only 50% is possible. The terms *high-density* and *low-density* polyethylene are thus used. Table 4.1 gives a comparison of the properties of the two forms. Low-density polyethylene softens in boiling water, the high-density does not. Low-density polyethylene has a lower tensile strength and lower tensile modulus than the high-density form. Both forms have excellent chemical resistance, low moisture absorption and high electrical resistivity. Low-density polyethylene is used mainly in the form of films and sheeting, e.g. for polythene bags, 'squeeze' bottles, ballpoint pen tubing, wire and cable insulation. High-density polyethylene is used for piping, toys, filaments for fabrics and household ware.

Polypropylene, in the form mainly used, has a linear chain with side groups regularly arranged along the chain (Figure 4.27). The presence of these side groups gives a more rigid and stronger polymer than

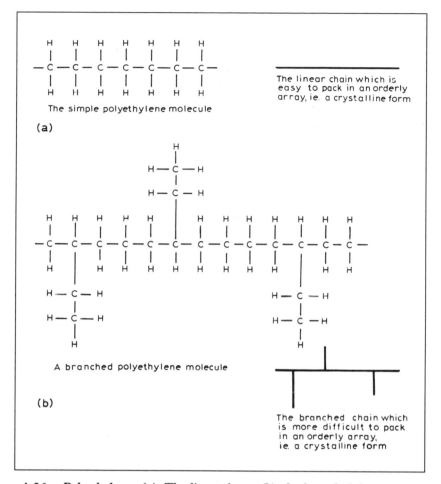

Figure 4.26   *Polyethylene. (a) The linear form; (b) the branched form*

Table 4.1   *Properties of polyethylene*

Property	Low density	High density
Crystallinity	60%	95%
Density (10^3 kg/m³)	0.92	0.95
Melting point (°C)	115	138
Tensile strength (MPa)	8 to 16	22 to 38
Tensile modulus (GPa)	0.1 to 0.3	0.4 to 1.3
% elongation	100 to 600	50 to 800
Max. service temperature (°C)	85	125

polyethylene in its linear form. The crystallinity is about 60%. Table 4.2 shows the properties. Polypropylene is used for crates, containers, fans, car fascia panels, tops of washing machines, cabinets for radios and televisions, toys and chair shells.

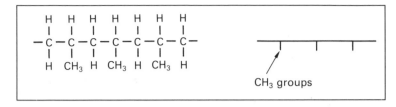

Figure 4.27 *Polypropylene*

Table 4.2 *Properties of polypropylene*

Property	
Crystallinity	60%
Density (10^3 kg/m³)	0.90
Melting point (°C)	176
Tensile strength (MPa)	30 to 40
Tensile modulus (GPa)	1.1 to 1.6
% elongation	50 to 600
Max. service temperature (°C)	150

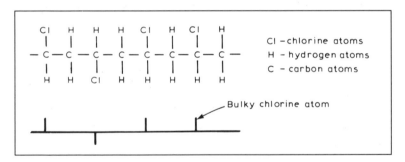

Figure 4.28 *PVC*

Polyvinyl chloride (PVC) has a linear chain but has 'bulky' atoms, chlorine atoms, on the chain (Figure 4.28). Because of this the chain is difficult to pack in an orderly manner and so gives a mainly amorphous material. When used without a plasticizer it is a rigid and relatively hard material. It is widely used for piping, but not for hot water as it has a maximum service temperature of 70°C. Above that temperature it softens too much. Most PVC products are, however, made with a plasticizer incorporated with the polymer. The amount of plasticizer is likely to be between about 5–50% of the plastic, the more plasticizer added, the greater the degree of flexibility. Table 4.3 shows the properties of PVC with different amounts of plasticizer. Plasticized PVC is used for the fabric of plastic raincoats, bottles, shoe soles, garden hose piping, gaskets and inflatable toys. All forms of PVC have good chemical resistance, though not as good as polythene.

Table 4.3   *Properties of PVC*

Property	No plasticizer	Low plasticizer	High plasticizer
Crystallinity	0%	0%	0%
Density (10^3 kg/m³)	1.4	1.3	1.2
Tensile strength (MPa)	52 to 58	28 to 42	14 to 21
Tensile modulus (GPa)	2.4 to 4.1		
% elongation	2 to 40	200 to 250	350 to 450
Max. service temperature (°C)	70	60 to 100	60 to 100

Table 4.4   *Properties of Polystyrene*

Property	No additives	Toughened
Crystallinity	0%	0%
Density (10^3 kg/m³)	1.1	1.1
Tensile strength (MPa)	35 to 60	17 to 42
Tensile modulus (GPa)	2.5 to 4.1	1.8 to 3.1
% elongation	1 to 3	8 to 50
Max. service temperature (°C)	65	75

Polystyrene has bulky side groups attached irregularly to the polymer chain and so gives an amorphous structure. Polystyrene with no additives is a brittle, transparent material with a maximum service temperature of about 65°C. It finds its main use as containers for cosmetics, light fittings, toys and boxes. A toughened form of polystyrene can be produced by blending polystyrene with rubber particles. This gives a marked improvement in properties, the material being less brittle. Table 4.4 shows the properties. This toughened material has a considerable number of uses, e.g. cups in vending machines, casings for cameras, projectors, radios, television sets and vacuum cleaners.

Acrylonitrile–butadiene–styrene terpolymer (ABS) is produced by forming polymer chains with three different polymer materials: polystyrene, acrylonitrile and butadiene. It gives an amorphous material which is tough, stiff and abrasion resistant. Table 4.5 shows the properties. ABS is widely used as the casing for telephones, vacuum cleaners, hair dryers, radios, television sets, typewriters, luggage, boat shells and food containers.

Polyamides, or nylons as they are better known, are linear polymers which can be designed to give different lengths of carbon–hydrogen chains between blocks of other atoms. There are a number of commonly used polyamides, e.g. nylon 6 and nylon 6.6. The numbers refer to the lengths of the molecular chains which are stuck together to give the polymer. Nylon 6 has lengths of a six-carbon long molecule stuck together (Figure 29(a)). Nylon 6.6 has two different six-carbon length molecules stuck together (Figure 2.29(b)). Table 4.6 shows the properties of these polymers. In general, nylon materials are strong, tough and have relatively

Table 4.5   *Properties of ABS*

Property	
Crystallinity	0%
Density (10^3 kg/m³)	1.1
Tensile strength (MPa)	17 to 58
Tensile modulus (GPa)	1.4 to 3.1
% elongation	10 to 140
Max. service temperature (°C)	110

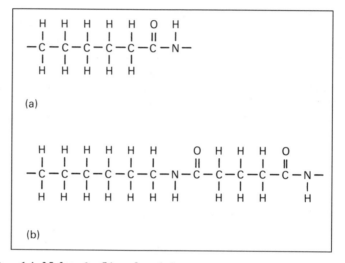

Figure 4.29   *(a) Nylon 6; (b) nylon 6.6*

Table 4.6   *Properties of nylons*

Property	Nylon 6	Nylon 6.6
Crystallinity	Can be varied from low to high percentages	
Density (10^3 kg/m³)	1.13	1.1
Melting point (°C)	225	265
Tensile strength (MPa)	75	80
Tensile modulus (GPa)	1.1 to 3.1	2.8 to 3.3
% elongation	60 to 320	60 to 300
Max. service temperature (°C)	110	110

high melting points. They do, however, tend to absorb moisture, the effect of which is to reduce their tensile strength. Nylon 6.6 can absorb quite large amounts of moisture. Nylons often contain additives, e.g. glass spheres or fibres, to give improved strength and stiffness, stabilizers and flame retardents. Molybdenum disulphide is used as an additive with nylon 6 to give a material with very low frictional properties. Nylons are used for the manufacture of fibres for clothing, gears, bearings, bushes, housings for domestic and power tools, electric plugs and sockets.

4.8 Thermosets

The atoms in a thermoset form a three-dimensional structure of chains with frequent cross-links between chains (Figure 4.21(c)). The bonds linking the chains are strong and not easily broken. Thus the chains cannot slide over one another. As a consequence, thermosetting polymers are stronger and stiffer than thermoplastics.

Thermoplastics offer the possibility of being heated and then pressed into the required shapes. Thermosets cannot be so manipulated. The processes by which thermosetting polymers can be shaped are limited to those where the product is formed by the chemicals being mixed together in a mould so that the cross-linked chains are produced while the material is in the mould. The result is a polymer shaped to the form dictated by the mould. No further processes, other than possibly some machining, are likely to occur.

4.8.1   Examples of thermosets

The following is a brief outline of the nature and properties of commonly used thermosets. Such materials are widely used with open-weave fabrics, such as glass-fibre fabric, to give composites (see Section 4.11).

Phenolics give highly cross-linked polymers. *Phenol formaldehyde* was the first synthetic plastic and is known as *Bakelite*. The polymer is opaque and initially light in colour. It does, however, darken with time and so is always mixed with dark pigments to give a dark-coloured material. It is supplied in the form of a moulding powder which includes the polymer, fillers and other additives such as pigments. When this moulding powder is heated in a mould the cross-linked polymer chain is produced. The fillers account for some 50–80% of the total weight of the moulding powder. Wood flour, a very fine softwood sawdust, when used as a filler increases the impact strength of the plastic, asbestos fibres improve the heat properties, and mica the electrical resistivity. Table 4.7 shows some of the properties of this thermoset. Phenol formaldehyde mouldings are used for electrical plugs and sockets, switches, door knobs and handles, camera bodies and ash trays. Composite materials involving the polymer being used with paper or an open-weave fabric, e.g. a glass fibre fabric, are used for gears, bearings and electrical insulation parts.

Amino-formaldehyde materials, generally *urea formaldehyde* and *melamine formaldehyde*, give highly cross-linked polymers. Both are used as moulding powders with cellulose and wood flour widely used as fillers. Hard, rigid, high-strength materials are produced. Table 4.8 shows some of the properties. Both materials are used for tableware (e.g. cups and saucers), knobs, handles, light fittings and toys. Composites with open-weave fabrics are used as building panels and for electrical equipment.

Table 4.7   *Properties of phenol formaldehyde thermosets*

Property	Unfilled	Wood flour filler	Asbestos filler
Density (10^3 kg/m³)	1.25 to 1.30	1.32 to 1.45	1.6 to 1.85
Tensile strength (MPa)	35 to 55	40 to 55	30 to 55
Tensile modulus (GPa)	5.2 to 7.0	5.5 to 8.0	0.1 to 11.5
% elongation	1 to 1.5	0.5 to 1	0.1 to 0.2
Max. service temperature (°C)	120	150	180

Table 4.8   *Properties of amino-formaldehyde thermosets*

Property	Urea formaldehyde cellulose filler	Melamine formaldehyde cellulose filler
Density (10^3 kg/m³)	1.5 to 1.6	1.5 to 1.6
Tensile strength (MPa)	50 to 80	55 to 85
Tensile modulus (GPa)	7.0 to 13.5	7.0 to 10.5
% elongation	0.5 to 1.0	0.5 to 1.0
Max. service temperature (°C)	80	95

Epoxide materials are thermosets which are generally used in conjunction with glass (or other) fibres to give hard and strong composites. *Polyesters* can be produced as either thermosets or thermoplastics. The thermoset form is mainly used with glass (or other) fibres to form hard and strong composites. Such composites are used for boat hulls, architectural panels, car bodies, panels in aircraft, and stackable chairs.

4.9 Elastomers

Elastomers are polymers which can show very large, reversible, strains when subject to stress. Figure 4.30 shows a typical stress–strain graph for an elastomer. As the graph indicates, there is no linear part to it and so Hooke's law is not obeyed. The behaviour of the material is perfectly elastic up to considerable strains, e.g. you can stretch a rubber band up to more than five times its unstrained length and it is still elastic.

Elastomers have an amorphous polymer structure, consisting of tangled chains which are held together by occasional cross-linked bonds. The difference between thermosets and elastomers is that with thermosets there are frequent cross-linking bonds between chains while with elastomers there are only occasional bonds. A simple model for the elastomer structure might be a piece of very open netting (Figure 4.31). In the unstretched state the netting is in a loose pile. In the elastomer there will be some weak, temporary attractive forces (termed Van der Waals' forces) between chains in close proximity to each other, these being responsible for holding the tangled chains together. When the material begins to be

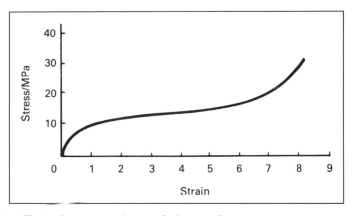

Figure 4.30   *Typical stress–strain graph for an elastomer*

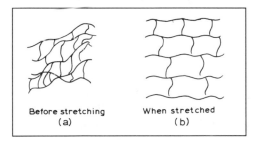

Figure 4.31   *Stretching an elastomer*

stretched the netting just begins to untangle itself and large strains can be produced. The weak, temporary forces between the chains will cause the elastomer to spring back to its original tangled state when the stretching forces are removed. It is not until quite large strains, when the netting has become fully untangled and the structure is orderly, that the bonds between atoms in the materials begin to be significantly stretched. At this point the stress–strain graph starts to become more steep and much larger stresses are needed to give further extensions.

4.9.1   Examples of elastomers

The following is a brief outline of elastomers, i.e. rubbers, which are commonly encountered in engineering. Table 4.9 gives some of the properties.

Natural rubber is, in its crude form, just the sap from a particular tree. The addition of sulphur to the rubber produces cross-links, the amount of cross-linkage being determined by the amount of sulphur added. The process of producing cross-links is termed *vulcanization*. Antioxidants and plasticisers are also added to the rubber.

An example of synthetic rubber is *butadiene styrene rubber*, commonly called SBR or GR-S or *Buna S* rubber. This is cheaper than natural rubber

Table 4.9 *Properties of elastomers*

Material	Tensile strength (MPa)	% elongation	Service temp. (°C)
Natural rubber	30	800	− 50 to + 80
Buna S	24	600	− 50 to + 80
Butyl rubber	20	900	− 50 to + 100
Nitrile rubber	28	700	− 50 to + 125
Neoprene	25	1000	− 50 to + 100
Polyurethane rubber	36	650	− 55 to + 125

and is used in the manufacture of tyres, hosepipes, conveyor belts and cable insulation. Another example is *butyl rubber*, often referred to as *isobutylene isoprene* or GR-I. This rubber has an important property of extreme impermeability to gases and thus is widely used for the inner linings of tubeless tyres, steam hoses and diaphragms. *Nitrile rubbers*, known as *butadiene acrylonitrile* or *Buna N*, are extremely resistant to organic liquids and are used for such applications as hoses, gaskets, seals, tank linings, rollers and valves. *Neoprene*, known as *polychloroprene*, has good resistance to oils and a variety of other chemicals, as well as good weathering characteristics. It is used for oil and petrol hoses, gaskets, seals, diaphragms and chemical tank linings. *Polyurethane rubbers* have higher tensile strengths, tear and abrasion resistance than other rubbers, are relatively hard and offer good resistance to oxygen and ozone. They are used for oil seals, diaphragms, tyres of forklift trucks and other vehicles where low speeds are involved (not high speeds since they have a low skid resistance), heels and soles of shoes and industrial chute linings.

4.10 The structure of composites

Composites are composed of two different materials bonded together in such a way that one serves as the matrix and surrounds fibres or particles of the other. A common example of a composite is reinforced concrete. This has steel rods embedded in concrete (Figure 4.32). The composite material carries loads that otherwise could not have been carried by the concrete alone.

There are many examples of composite materials encountered in every-day products. Many plastics are glass fibre or glass particle reinforced.

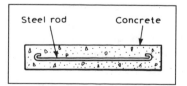

Figure 4.32 *Reinforced concrete*

Vehicle tyres are rubber reinforced with woven cords. Cermets, widely used for the tips of cutting tools, are composites involving ceramic particles in a metal matrix.

4.10.1 Fibres in a matrix

The fibres used in a matrix can be continuous long lengths all aligned parallel to an axis of the material, like the steels rods in reinforced concrete, or short fibres randomly orientated in the material. The long-length fibres give a directionality to the properties, the tensile strength and tensile modulus being much higher along the direction of the fibres than at right angles to them. Randomly orientated short fibres do not lead to this directionality of properties but do not offer such high tensile strengths or tensile modulus values. For example, a glass fibre with reinforced plastic (polyester) might have with long fibres a tensile strength of 800 MPa in the direction of the fibres and only 30 MPa at right angles to them. With short fibres the tensile strength in all directions might be 110 MPa.

Consider a composite rod made up of continuous fibres, all parallel to the rod axis, in a matrix (Figure 4.33). When tensile forces are applied to the composite rod, then each element in the composite has a share of the applied forces. Thus

Total force = forces on fibres + force on matrix

But since stress = force/area then the force on the fibres is equal to the product of the stress σ_f on the fibres and their total cross-sectional area A_f. Likewise, the force on the matrix is equal to the product of the stress σ_m on the matrix and its cross-sectional area A_m. Hence

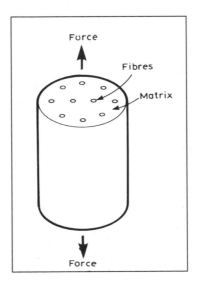

Figure 4.33 *Continuous fibres in a matrix*

Total force $= \sigma_f A_f + \sigma_m A_m$

Dividing both sides of the equation by the total area A of the composite gives

$$\text{Stress on composite} = \frac{\text{total force}}{\text{total area}}$$

$$= \sigma_f \frac{A_f}{A} + \sigma_m \frac{A_m}{A}$$

Thus the stress on the composite is the stress on the fibres multiplied by the fraction of the area that is fibres plus the stress on the matrix multiplied by the fraction of the area that is matrix.

Suppose we have glass fibres with a tensile strength of 1500 MPa in a matrix of polyester with a tensile strength of 45 MPa. If the fibres occupy, say, 60% of the cross-sectional area of the composite then the above equation indicates that the tensile strength of the composite, i.e. the stress the composite can withstand when both the fibres and matrix are stressed to their limits, will be

Strength of composite $= 1500 \times 0.6 + 45 \times 0.4 = 918$ MPa

If the fibres are firmly bonded to the matrix, then the elongation of the fibres and matrix must be the same and equal to that of the composite as a whole. Thus

Strain on composite = strain on fibres = strain on matrix

Dividing the stress equation above by this strain gives, since stress/strain is the tensile modulus,

$$\text{Modulus of composite} = E_f \frac{A_f}{A} + E_m \frac{A_m}{A}$$

Suppose we have glass fibres with a tensile modulus of 76 GPa in a matrix of polyester having a tensile modulus of 3 GPa. If the fibres occupy, say, 60% of the cross-sectional area of the composite then the tensile modulus of the composite is

Modulus of composite $= 76 \times 0.6 + 3 \times 0.4 = 46.8$ GPa

Example
A column of reinforced concrete has steel reinforcing rods running through its entire length and parallel to its axis. If the concrete has a modulus of elasticity of 20 GPa and the steel 210 GPa, what is the modulus of elasticity of the column if the steel rods occupy 10% of the cross-sectional area? Using the above equation

Modulus of composite $= 210 \times 0.1 + 20 \times 0.9 = 39$ GPa

Example

Carbon fibres with a tensile modulus of 400 GPa are used to reinforce aluminium with a tensile modulus of 70 GPa. If the fibres are long and parallel to the axis along which the load is applied, what is the tensile modulus of the composite when the fibres occupy 50% of the composite area?

Using the above equation

Modulus of composite $= 400 \times 0.5 + 70 \times 0.5 = 235$ GPa

4.11 Electrical conductivity

In terms of their electrical conductivity, materials can be grouped into three categories, namely conductors, semiconductors and insulators. Conductors have electrical conductivities of the order of 10^6 S/m, semiconductors about 1 S/m and insulators 10^{-10} S/m. Conductors are metals with insulators being polymers or ceramics. Semiconductors include silicon, germanium, and compounds such as gallium arsenide. Silicon is the most widely used semiconductor.

In discussing electrical conduction in materials, it is useful to visualize an atom as consisting of a nucleus surrounded by its electrons. The electrons are bound to the nucleus by electric forces of attraction. The force of attraction is weaker the further an electron is from the nucleus. The electrons furthest from the nucleus are called the valence electrons since they are the ones involved in the bonding of atoms together to form compounds.

Metals can be considered to have a stucture of atoms with valence electrons which are so loosely attached that they drift off and can move freely between the atoms. Thus, when a potential difference is applied across a metal, there are large numbers of free electrons able to respond and give rise to a current. We can think of the electrons pursuing a zigzag path through the metal as they bounce back and forth between atoms. An increase in the temperature of a metal results in a decrease in the conductivity. This is because the temperature rise does not result in the release of any more electrons but causes the atoms to vibrate and scatter the electrons more.

Insulators, however, have a structure in which all the electrons are tightly bound to atoms. Thus, there is no current when a potential difference is applied because there are no free electrons able to move through the material. To give a current, sufficient energy needs to be supplied to break the strong bonds which exist between electrons and insulator atoms. The bonds are too strong to be easily broken and hence there is no current. A very large temperature increase would be necessary to shake such electrons from the atoms.

Semiconductors can be regarded as being insulators at a temperature of absolute zero. However, the energy needed to remove an electron from an atom is not very high and at room temperature there can be sufficient energy supplied for some electrons to break free. Thus, the application of a potential difference will result in a current. Increasing the temperature results in more electrons being shaken free and, hence, an increase in conductivity. When a silicon atom looses an electron it is an electron short and we can consider there to be a hole in its valence electrons. When electrons move they can be thought of as hopping from valence site to a hole in a neighbouring atom, then being released and moving to another hole, etc. Not only do electrons move through the material but so do the holes.

The conductivity of a semiconductor can be very markedly changed by impurities. For this reason the purity of semiconductors must be very carefully controlled. The impurity level of silicon used for the manufacture of semiconductor devices is routinely controlled to less than one atom in a thousand million silicon atoms. Foreign atoms can however be deliberately introduced in controlled amounts into a semiconductor in order to change its electrical properties. This is referred to as *doping*. Atoms such as phosphorus, arsenic or antimony, when added to silicon, add easily released electrons and so make more electrons available for conduction. Such dopants are called *donors*. Semiconductors with more electrons available for conduction than holes are called an *n-type semiconductor*. Atoms such as boron, gallium, indium or aluminium add holes into which electrons can move. They are thus referred to as *acceptors*. Semiconductors with an excess of holes are called a *p-type semiconductor*.

Problems

1 Explain the term *grain* when used in connection with the structure of metals.
2 Explain what is meant by the term *alloy*.
3 Explain the terms *ferrous alloy* and *non-ferrous alloy*.
4 Describe the structure of metals.
5 How does the grain size in a metal affect its properties?
6 How does the shape of grains within a metal affect its properties?
7 Describe the effects on the grain structure and properties of a metal of cold working.
8 Describe the effects on the properties of carbon steels of increasing the percentage of carbon in the alloy.
9 What types of structure might you expect for a metal which is (a) ductile and (b) brittle?
10 A pure metal is formed into an alloy by larger atoms being forced into the spaces in its crystal structure. What changes might be expected in the properties and why?
11 Describe how the mechanical properties of a cold-worked metal changes as its temperature is raised from room temperature to about $0.6T_m$, where T_m is the melting point on the kelvin scale.
12 How does the temperature at which working is carried out determine the grain size and so the mechanical properties?
13 Why are the mechanical properties of a cold-rolled metal different in the direction of rolling from those at right angles to this direction?
14 How does a cold-rolled product differ from a hot-rolled one?

15 Brasses have recrystallization temperatures of the order of 400°C. Roughly, what temperature should be used for the hot extrusion of brass?

16 A brass, 65% copper and 35% zinc, has a recrystallization temperature of 300°C after being cold worked so that the cross-sectional area has been reduced by 40%.
(a) How will further cold working change the structure and properties of the brass? (b) To what temperature should the brass be heated to give stress relief? (c) To what temperature should the brass be heated to anneal it?

17 Use Figure 4.16 for this problem. According to this Figure:
(a) What is the maximum hardness possible with cold-rolled copper? (b) Copper plate, already cold worked 10%, is further cold worked 20%. By approximately how much will the hardness change? (c) Mild steel is to be rolled to give thin sheets. This involves a 70% reduction in sheet thickness. What treatment would be suitable to give this reduction and a final product which was no harder than 150 HV?

18 Describe the difference between amorphous and crystalline polymer structures and explain how the amount of crystallinity affects the mechanical properties of the polymer.

19 Compare the properties of low- and high-density polyethylene and explain the differences in terms of structural differences between the two forms.

20 Why are (a) stabilizers; (b) plasticizers and (c) fillers added to polymers?

21 Describe how the properties of PVC depend on the amount of plasticizer present in the plastic.

22 Increasing the amount of sulphur in a rubber increases the amount of cross-linking between the molecular chains. How does this change the properties of the rubber?

23 Explain how elastomers can be stretched to several times their length and still be elastic and return to their original length.

24 Calculate the tensile modulus of a composite consisting of 45% by volume of long aligned glass fibres, tensile modulus 76 GPa, in a polyester matrix, tensile modulus 4 GPa. In what direction does your answer give the modulus?

25 In place of the glass fibres referred to in problem 24, carbon fibres are used. What would be the tensile modulus of the composite if the carbon fibres had a tensile modulus of 400 GPa?

26 Long boron fibres, tensile modulus 340 GPa, are used to make a composite with aluminium as the matrix, the aluminium having a tensile modulus of 70 GPa. What would be the tensile modulus of the composite in the direction of the aligned fibres if they constitute 50% of the volume of the composite?

27 How will the properties of composites differ if they are (a) made of long fibres all orientated in the same direction and (b) short fibres with random orientation?

5 Processing of materials

> **Outcomes** At the end of this chapter you should be able to:
> - Identify the processing routes for particular materials in order to produce particular microstructures.
> - Identify the changes in properties and microstructure associated with processing routes.

5.1 Shaping metals

The main methods used to shape metals are:

1 Casting, in which a product is formed by pouring liquid metal into a mould. Sand casting involves using a mould made of sand, die casting uses a metal mould.
2 Manipulative processes, in which a shape is produced by plastic deformation processes. This includes such cold-working methods as rolling, drawing, pressing and impact extrusion. Hot-working processes include rolling, forging and extrusion.
3 Powder techniques, in which a shape is produced by compacting a powder.
4 Cutting and grinding, in which a shape is produced by metal removal.

The above shaping processes are one way of producing a product. Another method is metal joining, of which the main processes are:

1 Adhesives
2 Soldering and brazing
3 Welding
4 Various fastening systems, e.g. rivets, bolts and nuts.

The shaping or assembly method used for a particular product will depend on the metal to be used, its form of supply, and the form of the product. The commercial forms of supply of materials might be as bars, sheet, ingot or pellet.

This chapter is not about the details of the various shaping methods that can be used with either metals or polymers but the structural changes that occur in materials as a result of such processes.

5.1.1 Casting

Casting, in the case of metals, is the shaping of an object by pouring the liquid metal into a mould and then allowing it to solidify and form a product with the internal shape of the mould. The metal has to flow into all parts of the mould. Think of the problems of pouring treacle into a mould compared with pouring water. With the water it is much easier to fill all parts of the mould. Thus alloys used for casting where gravity is used to get the liquid metal to flow, as in sand casting and gravity die casting, have their alloying constituents chosen to give good flowing properties, i.e. low viscosity. Where pressure is used to force the liquid metal into the mould, as in pressure die casting, then a more viscous alloy can be used.

The grain structure within the product is determined by the rate of cooling. Figure 5.1 (see Figure 4.4 for a photograph) shows the grain structure in a casting. Where the cooling rate is high then small grains are produced, where the rate is low then larger grains are produced. Thus, since the metal in contact with the mould cools faster than that in the centre of the casting, smaller grains are produced at the mould surfaces than in the centre. The small grains produced near the surface are called *chill crystals*. The cooling rate a little in from the mould walls is less than that at the walls. A consequence of this is that some chill crystals, given time, can develop into long elongated crystals in an inward direction. Such grains are called *columnar crystals*. The grains produced in the centre of the casting where the cooling rate is the slowest are called *equiaxed crystals*. These crystals grow in liquid metal which is constantly on the move due to convection currents. As a consequence, the crystals are almost spherical.

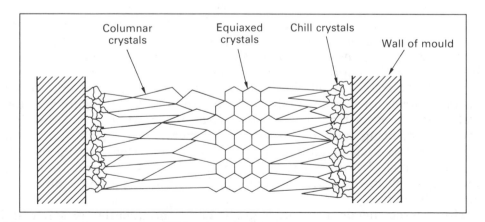

Figure 5.1   *Structure on solidification of an ingot*

Castings in which the mould is made of sand tend to have a slow rate of cooling as sand has a low thermal conductivity. Thus sand castings tend to have large columnar grains. Since large grains mean a low strength and hardness then sand castings have relatively low strength and hardness. Die castings involving metal moulds have a much faster rate of cooling and so give castings for which there has not been enough time for long columnar crystals to develop and therefore have a bigger zone of equiaxed crystals. As these are relatively small crystals then the casting has better mechanical properties than the corresponding sand casting. For example, an aluminium alloy casting alloy LM6 gives, when sand cast, a tensile strength of about 160 MPa and when die cast 190 MPa, the corresponding elongations being 5% and 7%.

Example
What type of grain structure might be expected when liquid metal is poured into a narrow metal mould?

The rate of cooling will be high because it is a metal mould. Also, because the mould is thin, all parts of the casting will cool quickly. Thus it is likely that chill crystals will be formed throughout since there is not enough time for columnar crystals to develop.

5.1.2 Manipulative processes

Manipulative methods involve the shaping of a material by plastic deformation processes. The products given by such methods are said to be *wrought*. Where the deformation is carried out below the recrystallization temperature, the process is said to involve *cold working*, when in excess of that temperature *hot working* (see Sections 4.4 and 4.5). The main cold-working processes are cold rolling, drawing, pressing and impact extrusion. The main hot-working processes are rolling, forging and extrusion.

Cold-worked metals generally have deformed grains and consequently are harder and more brittle, being said to be work hardened. They also have a directionality of properties since the grains are deformed in the direction of the manipulation. Thus with cold rolling, the sheet in the direction of the rolling has different properties at right angles to that direction. Cold rolling may have to take place in a number of stages with annealing between the stages. This is because the metal becomes too work hardened for the entire reduction in thickness to occur with one operation. The annealing results in the grains becoming large and undistorted again. *Drawing* involves the pulling of metal through a die, Figure 5.2 illustrating this process for wire drawing. *Deep drawing* (Figure 5.3) involves sheet metal being pushed through an aperture by a punch. The more ductile metals such as aluminium, brass and mild steel can be used to shape products by this method. Both the drawing or wires and deep drawing result in the material work hardening and assuming a directionality of properties.

Hot-worked metals have larger grains and consequently are softer and more ductile. Hot rolling is generally at a temperature of about $0.6T_m$,

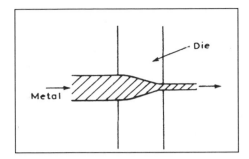

Figure 5.2 *Drawing a wire*

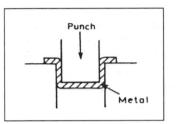

Figure 5.3 *Deep drawing*

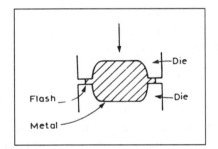

Figure 5.4 *Closed-die forging*

where T_m is the melting point of the metal on the kelvin scale. At this temperature work hardening does not occur though the surfaces of the material become oxidized. *Forging* involves squeezing, either by pressing or hammering, a ductile metal between a pair of dies so that it ends up assuming the internal shape of the dies. Figure 5.4 illustrates this process in relation to what is termed *closed-die forging*. Generally, for the material to be ductile enough during the operation the forging has to take place at a temperature in excess of the recrystallization temperature. The flow of the material during the squeezing operation does give some directionality to the properties of the material with grains and any non-metallic inclusions in the metal tending to end up aligned along the directions of flow.

Extrusion is rather similar to the squeezing of toothpaste out of a tube. The shape of the extruded toothpaste is determined by the nozzle

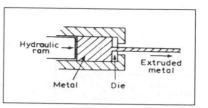

Figure 5.5 *Extrusion*

through which it is ejected. Figure 5.5 shows the basic principles. Cold extrusion is when the extrusion takes place below the recrystallization temperature, hot extrusion at a higher temperature. Typically, hot extrusion takes place at a temperature of the order of 0.65 to $0.9T_m$. Cold extrusion gives a work-hardened product, hot extrusion a soft, ductile, one. Hot extrusion is generally required when the cross-section of the material is reduced by a factor of 50 or more. The flow of the material that occurs during extrusion ends up in giving some directionality of grains and hence properties.

Example
What types of carbon steels can be used with the deep drawing process?

For deep drawing high ductility is required. As a consequence, only carbon steels with low amounts of carbon can be used as only they have sufficient ductility (see Figure 4.7 and the associated text). Thus mild steel can be used but not a high-carbon steel.

5.1.3 Powder techniques

Shaped metal components can be produced from a metal powder. The process, called *sintering*, involves compacting the powder in a die and then heating it to a sufficiently high temperature for the particles to knit together. The compacting forces particles into contact with each other and then the heating enables atoms in particles to diffuse across the points of contact and form necks which hold the particles together when the material cools.

Sintering is a useful method for the production of components from brittle materials where manipulative processes are difficult and high melting point materials for which melting for casting becomes too expensive.

5.2 Shaping polymers

Polymers may be supplied in a powder, granule or sheet form, the supplier having mixed the polymer with suitable additives and even other polymers in order that, after processing, the finished material should have the required properties. The main processes used to shape polymers are:

1   Casting. This can involve the mixing of the constituent parts of the plastic in a mould and then allowing the resulting chemical reaction to produce the polymer. This method can be used with thermosets and thermoplastics. Another method involves the melting of the powdered polymer in a heated mould.

2   Moulding. With thermoplastics the polymer might be melted and forced into a mould, the process being called *injection moulding*. With thermosets, the powdered polymer may be compressed between the two parts of the mould and then heated under pressure. This process is known as *compression moulding*. With *transfer moulding* the powdered thermoset is heated in a chamber before being transferred by a plunger into the mould.

3   Forming, in which sheet polymer, a thermoplastic, is heated and pressed into or around a mould.

4   Extrusion, in which a thermoplastic polymer is forced through a die.

In addition, products may be formed by polymer joining. The main processes are:

1   Adhesives
2   Welding
3   Various forms of fastening systems, e.g. riveting, press and snap fits, screws.

5.2.1   Flowing processes

Many of the polymer processes used to produce products can be considered to involve the flowing of liquid polymer and its subsequent cooling in the required shape. Figure 5.6 shows the basic principle of one form of *injection moulding,* a very widely used process. The polymer is melted and forced into a mould. The method is capable of giving high production rates because the rate of cooling of the polymer in the mould is fairly fast.

Consider a linear chain thermoplastic material, which is capable of crystallizing. While the polymer is in the liquid state the molecules are able to move about. For chains of the polymer to form there has to be sufficient time for the molecules being incorporated into the chain to arrive at the correct positions in the chain. Thus starting with a small length of

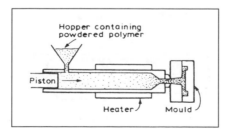

Figure 5.6   *Injection moulding*

chain, we can think of the chain steadily increasing in length as more and more molecules arrive at its ends and join on. We can consider that some of these chains, while growing, fold to give a crystal. These crystals can then grow as more molecules attach themselves to the ends of the chain. All this, however, takes time. Thus, the size of the polymer crystals, indeed whether there is crystallinity at all, depends very much on the time during which the polymer is liquid. Hence it depends on the rate of cooling. Large crystals can be produced by slow cooling, small crystals by fast cooling. If there is very fast cooling then no crystallinity may occur and the polymer is completely amorphous.

Because injection moulding has a fast cooling rate, necessary for high production rates, it can result in a polymer, which could be crystalline at slow rates of cooling, being completely amorphous. This has consequences for the properties of the material. Amorphous polymers can be stretched more than crystalline polymers. For some polymers, the rate of cooling is slow enough for the inner part of the moulded shape to crystallize while the outer layers, which cool faster than the inner part, are amorphous. As a consequence, because crystalline polymer chains are more tightly packed than randomly arranged ones, the inner part has a higher density than the outer layers.

With injection moulding, liquid polymer is forced into the mould. The direction of flow of the polymer usually varies with position throughout the mould. Since the molecules tend to align with the direction of the flow, the plastic product can have mechanical properties which vary with direction.

Extrusion involves the forcing of liquid polymer, a thermoplastic, through a die. The process is comparable with the squeezing of toothpaste out of its tube. Figure 5.7 shows the basic form of the extrusion process. The polymer is fed into a screw mechanism which takes it through the heated zone and forces it out through the die. Here it cools rapidly. As a consequence, the cooling might be too fast for crystals to develop, so an amorphous polymer product results. In some cases the inner core of the extruded polymer might cool slow enough for some crystallinity, while just the outer layers remain amorphous. The direction of the polymer chains is also likely to follow that of flow of the polymer and so give a product with different properties along the length of the extrusion than at right angles.

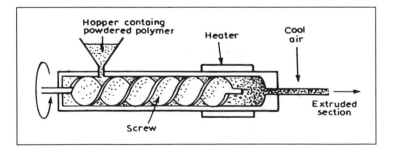

Figure 5.7 *Extrusion*

Example
One way of producing sheet plastic is to extrude the polymer through a slit. What types of properties might be expected of the sheet?

The extrusion process will tend to align the polymer molecules with the direction of flow. Thus the sheet might be expected to have different properties in directions along the length of the sheet and at right angles to it.

5.2.2 Manipulative processes

Forming processes are used to form articles from sheet thermoplastic polymer. The heated sheet is pressed into or around a mould. The term *thermoforming* is often used. The sheet may be pressed against the mould by the application of pressure on the sheet, this method being called thermoforming. Alternatively, it can be by the production of a drop in pressure between the sheet and the mould, as illustrated in Figure 5.8. This method is called *vacuum forming*.

Forming processes involve a polymer being stretched. As a consequence, the molecular chains are forced into becoming aligned in the direction of the stretching (see Section 4.7). Therefore the formed product has a directionality of properties.

5.2.3 Drawing of polymers

Stretching a thermoplastic polymer can result in molecular chains becoming lined up in the direction of the applied forces (see Section 4.7). The polymer with its molecules all thus orientated has different mechanical properties from the non-orientated polymer, being considerably stronger. Thus cold stretching such a polymer can improve its strength. The process is generally termed *cold drawing*. However, if the polymer is stretched while hot and then the forces removed it is still possible for the molecules to become disorientated during the cooling and no such alignment of molecules and improvement in properties occurs. Hot stretching has virtually no effect on the properties.

The plastic bottles used for fizzy drinks, e.g. Coca-Cola, are made from polyethylene terephthalate (PET). The method uses injection moulding to

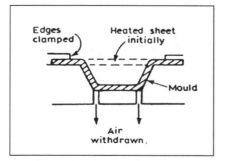

Figure 5.8 *Vacuum moulding*

produce an initial bottle-shape. If the PET material is cooled quickly after the moulding then the material is amorphous, if cooled slowly crystalline. The amorphous material is transparent, permeable to the 'fizz' and not very strong. The crystalline material is opaque, impermeable and stronger. What is wanted, however, is the combination of transparency with impermeability and strength. The procedure that is adopted is to first obtain the amorphous form. Then the bottle is heated, but not so high that plastic flow can occur, and stretched in length and generally expanded to the required shape by air pressure blowing the bottle out to its full size in another mould. This results in some molecular alignment occurring with a consequent improvement in properties while still retaining the transparency.

Example
Nylon thread is often cold drawn after its production. What effect is this likely to have on the structure and properties of the nylon?

The cold drawing improves the alignment of the polymer molecules and so results in improved strength.

5.3   Heat treatment of metals

Heat treatment can be defined as the controlled heating and cooling of metals in the solid state for the purpose of altering their properties. A heat treatment cycle consists normally of three parts:

1   Heating the metal to the required temperature for the changes in structure within the material to occur.
2   Holding at that temperature for long enough for the entire material to reach the required temperature and the structural changes to occur throughout the entire material.
3   Cooling, with the rate of cooling being controlled since it affects the structure and hence properties of the material.

Annealing is the heat treatment used to make a metal softer and more ductile. It involves heating the metal to above the recrystallization temperature (see Section 4.4.1) and then slow cooling. The result is a regrowth of grains to give a large grain structure.

In the case of carbon steels, this change in grain size is also accompanied by changes in the form of the constituents present in the alloy. Heating a steel to above the recrystallization temperature changes the crystal structure from ferrite to austenite (see Section 4.2.1). The austenite can contain more carbon atoms than the ferrite. There is time with annealing, because the cooling is slow, for the excess carbon atoms to move out of the crystal structure to form a compound called *cementite*. If the heated metal is cooled very rapidly, for example by being dropped into cold water, there is not time for the austenite to lose its excess carbon atoms and they become trapped in the structure. The result is a new structure called *martensite*. The carbon trapped in this structure considerably distorts the structure. As a

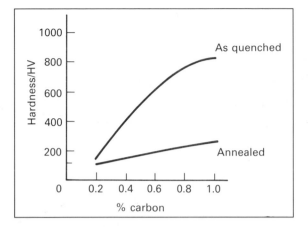

Figure 5.9 *Hardness of carbon steels*

consequence, martensite is very hard and brittle and the steel becomes harder and more brittle. The hardness and strength increases quite significantly with an increase in carbon content, this being because more carbon is trapped in the structure and so there is more distortion.

This form of heat treatment is called *quenching*. A problem with the severe cooling occurring in quenching is that cracks can occur. These are a result of the distortion produced by the structural changes and also differential expansion as a result of different parts of a product cooling at different rates. Figure 5.9 shows how the hardness of carbon steels after quenching compares with that of the annealed steels.

In the quenched state steels have such a low ductility as to be very difficult to use. The process known as *tempering* can, however, be used to improve the ductility without losing all the hardness gained by the quenching. Tempering involves heating the steel to a temperature at which some of the carbon trapped in the martensite structure can diffuse out and form cementite, so reducing the distortion of the structure. The amount of carbon that diffuses out depends on the temperature used for the tempering. Thus the mechanical properties depend on the tempering temperature. Figure 5.10 shows this for an alloy steel (a manganese–nickel–chromium–molybdenum steel).

A wide range of alloys used in engineering depend on a treatment called *precipitation hardening* for improvements in their hardness and strength. This type of treatment is widely used with aluminium alloys and nickel alloys. The process involves heating the alloy to above the recrystallization temperature, then quenching. The result is a distorted crystal structure. However, with time, atoms diffuse out of the structure of this type of alloy to give a fine precipitate. This precipitate lodges at grain boundaries and in dislocations and, as a consequence, makes slip more difficult. The result is an increase in hardness and strength. For example, the aluminium–copper alloy (2014) might have a tensile strength of 185 MPa and hardness 45 HB in the annealed state and, after precipitation hardening, 425 MPa and 105 HB.

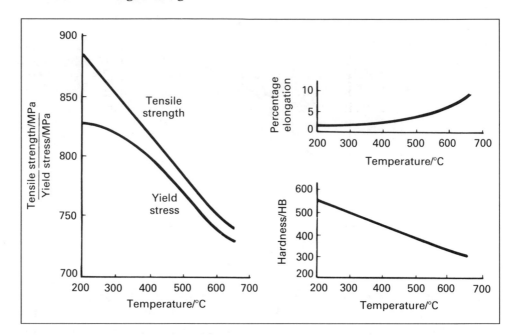

Figure 5.10   *The effect of tempering*

Example

The following are mechanical properties of a nickel alloy. Identify the internal structure which is responsible for the properties.

	Strength (MPa)	Yield stress (MPa)	Elongation (%)
Cold worked	535	380	12
Cold worked and annealed	380	105	36

The cold working has caused grains to be become distorted and include a greater number of dislocations. As a consequence, the material is strong and relatively brittle. Annealing the material has removed the distortion, increasing the grain size and reducing the number of dislocations. As a result, the material is weaker and more ductile.

Example

The following are the mechanical properties of a carbon steel. Identify the internal structure which is responsible for these properties.

	Strength (MPa)	Hardness (BH)	Elongation (%)
As rolled	815	240	17
Annealed	625	180	23
Quenched and tempered at 200°C	1100	320	13
Quenched and tempered at 650°C	800	230	23

The as-rolled material will have some work hardening and therefore distorted grains with increased numbers of dislocations. Annealing the material results in recrystallization with the result that large grains are produced with a drop in the number of dislocations. Therefore the material is weaker, softer and more ductile. Quenching results in the formation of martensite in which carbon atoms are trapped within a distorted structure. As a consequence, the material is much harder and more brittle. Tempering allows some of the carbon atoms to diffuse out and reduce the distortion, the higher the tempering temperature, the greater the reduction. Hence tempering restores some of the ductility but reduces the strength.

5.3.1   Surface hardening

There is often a need for the surface of a piece of steel to be hard, e.g. to make it wear resistant, without the entire component being made hard and often too brittle. Several methods are available for surface hardening.

For carbon steels surface hardening can be achieved by just heating the surface layers to above the recrystallization temperature and then quenching to give a martensitic structure for these surface layers. This selective heating of the surface layers can be carried out with an oxyacetylene flame, called *flame hardening*, or by placing the steel component in a coil carrying a high-frequency current and allowing the induced currents in the surface layers to do the heating, called *induction heating*. Another method that can be used to produce martensite in the surface layers is to increase the carbon content of the surface layers, this method being known as *case hardening*. This can be done by heating the steel while it is packed in charcoal and barium carbonate (*pack carburizing*) or in a furnace in an atmosphere of a carbon-rich gas (*gas carburizing*) or, alternatively, in a bath of liquid sodium cyanide (*cyanide carburizing*). These methods might, for example, result in a steel having an inner core containing 0.2% carbon and surface layers with 0.9% carbon.

Other processes that can be used involve changing the surface composition by diffusing nitrogen into it to produce hard compounds (nitrides). This is done by heating the steel in an atmosphere of ammonia gas and hydrogen. The process is known as *nitriding*. *Carbonitriding* involves heating the steel in an atmosphere containing both carbon and ammonia and allowing both carbon and nitrogen to diffuse into the surface layers.

5.4 Integrated circuit fabrication

A typical integrated circuit is a chip about 5 mm across by 1 mm thick and has been cut from a wafer of semiconductor material which contains hundreds of such chips. An integrated circuit contains a complete circuit made up from such components as transistors, diodes, resistors and capacitors. The components are fabricated by selective doping of the surface layers of the chip, the components being formed in the top 10 to 20 μm. The substrate is a p-type semiconductor with the components in the surface layers formed by the selective introduction of n- and p-type dopants. The surface of the chip is covered with silicon dioxide, which is an electrical insulator, with connections to the underlying components being made by aluminium which is deposited over the silicon dioxide and through holes (termed windows) in this layer.

The layer of silicon dioxide can be produced by a chemical reaction between the wafer surface and either oxygen or steam, or by a deposition process. Etching is used to remove unwanted regions of material from a wafer, e.g. cutting windows in the silicon dioxide layer through which doping of the underlying layers can occur. This can be done by:

1 Wet etching, in which wafers are immersed in an acid bath
2 Plasma etching, with the surface being exposed to reactive gas atoms, generated by the breakdown of a gas, as a result of being heated by radio-frequency electromagnetic energy
3 Ion milling, in which a beam of high energy ions are used as a bombarding beam to dislodge atoms.

Dopants can be introduced, through windows, into a wafer by solid diffusion or ion implantation. With solid diffusion, the dopant atoms are deposited on the surface and the wafer is then heated. The atoms gradually diffuse into the surface layers. With ion implantation, the dopant is fired at the wafer as a beam of ions.

Problems

1 Explain why copper wires that are used for electrical conductors are usually finished by a cold-drawing process and then heated to about 700°C.
2 What microstructure, and hence what properties, would you expect in cold-drawn wire if there is no further treatment of it?
3 Which type of casting, sand or die casting, will produce a product with the smallest grains?
4 With a thermoplastic polymer, how does the rate of cooling from the liquid state affect the degree of crystallinity?
5 What would you expect to be the internal structure of an extruded thermoplastic?
6 The thin plastic containers used to hold biscuits or chocolates in boxes are produced by thermoforming a thermoplastic. What is likely to be the resulting molecular structure in the various parts of the container?
7 Polythene bags are generally made by extruding the polyethylene through a die to give an extruded cylinder. The cylinder, while hot, is then inflated by air pressure. What type of structure might be expected within the bag material? You might care to try an experiment of cutting strips in directions along the length of the bag and at right angles and pulling them between your hands. The results can then be compared with the structure predicted by your answer.

8 Polypropylene twine consists of fibres of polypropylene which have been cold drawn. What is the effect of this process?

9 State what structural changes take place, and the consequential changes in properties: in (a) annealing; (b) quenching; (c) tempering; (d) precipitation hardening; (e) flame hardening and (f) case hardening.

10 A carbon steel is found to have the following properties. Explain how they arise in terms of the structure of the steel.

	Strength (MPa)	Hardness (HB)	Elongation (%)
As rolled	550	180	32
Annealed	465	125	32
Quenched, tempered at 200°C	850	495	17
Quenched, tempered at 650°C	585	210	32

11 An aluminium–manganese alloy is found to have the following properties. Explain how they arise in terms of the structure of the alloy.

	Strength (MPa)	Hardness (BH)	Elongation (%)
Annealed	110	28	30
Fully work hardened	200	55	4

12 An aluminium–magnesium–silicon alloy is found to have the following properties. Explain how they arise in terms of the structure of the alloy.

	Strength (MPa)	Hardness (BH)	Elongation (%)
Annealed	125	30	25
Precipitation hardened	310	95	12

13 The striking part of the head of a hammer is required to be very hard, but the main body of the hammer head must be softer and more tough. How can these properties be achieved in a single piece of steel?

14 What properties are required of a hacksaw blade and how might they be obtained?

6 | Selection of materials

Outcomes At the end of this chapter you should be able to:

* Identify the materials properties required for a particular specification/application.
* Recognize materials with the required properties.
* Suggest processing routes.
* Discuss the behaviour of materials in service.

6.1 The requirements

What functions does a product have to perform? This is an important question that requires an answer before either the materials or the forming processes for the materials are considered. From this stems a sequence of further questions. The following example serves to illustrate how the sequence might develop.

Consider the problem of making a domestic kitchen pan. Its functions may be deemed to be – to hold liquid and allow it to be heated to temperatures of the order of $100°C$. From a consideration of the function we can arrive at the basic design requirements. Thus, a consequence of these functions for the pan are the requirements for a particular shape of container which must not deform when heated to these temperatures. It must be a good conductor of heat. It must be leakproof. It must not ignite when in contact with a flame or hot electrical element. In addition, there may be other requirements which are not so essential, but certainly desirable. For the pan we might thus require an attractive finish.

From these requirements we can now define the required properties of the materials. Thus for the pan, the requirement that the material be a good conductor of heat would seem to reduce the consideration to metals, particularly when taken together with the requirement that the material can be put in contact with a flame and contain hot liquids. This would effectively rule out polymers. But what properties are required of the metal? The shape of the pan would suggest that a deep-drawing process be used (see Section 5.1.2). As this is a cold-working process then there will be a good surface finish. For deep drawing the initial material must be reasonably ductile and available in sheet form. Thus we might consider an aluminium alloy. If we look up tables, a possibility would seem to be an aluminium–magnesium alloy, AA5005. Another possibility would be a stainless steel, stainless because rusty pans would not be very desirable. If we look up tables, a possibility would seem to be 302S31. The deciding factor is likely to be cost, though there may be some prestige value attached to having stainless steel plans as opposed to aluminium, which would allow a higher price to be charged. For the same volume of material, the stainless steel will probably cost about three times the aluminium alloy.

The above represents one line of argument regarding the design of pans. It is instructive to examine a range of pans and consider the materials used and what reasons might be advanced for them being chosen. Why, for example, are some pans made of glass, of a ceramic, of a steel coated on the outside with an enamel and on the inside with a non-stick polymer polytetrafluoroethylene (PTFE)?

The above is only the consideration of the container part of the pan. There is still the handle to study. The function required is that it can be used to lift the pan and contents, even when they are hot. The properties required are thus poor thermal conductivity, able to withstand the temperatures of the hot pan, stiffness and adequate strength. The handle can be considered to be essentially a cantilever with a load, the pan and contents, at its free end. Before going too far in considering the design and materials for the handle, British Standards should be consulted. BS 6743 gives a standard specification for the performance of handles and handle assemblies attached to cookware. This sets the levels of safe performance against identified tests simulating hazards experienced in normal service. Taking this into consideration, then the need for the handle to have low thermal conductivity indicates that metal would not be a good choice. The possibility is thus a polymer. It needs, however, to be able to withstand a temperature of the order of 100°C at the pan end and have a reasonably high modulus of elasticity and reasonable strength. These requirements suggest that a thermoset is more likely to be feasible than a thermoplastic. A possibility is phenol formaldehyde (Bakelite). The dark colour of this material is no problem in these circumstances. When filled with, say, wood flour it has a high enough maximum service temperature of about 150°C, a tensile modulus of 5.0 to 8.0 GPa (high for a polymer), and a tensile strength of 40 to 55 MPa (see Table 4.7). Because it is a thermoset then the processing method could be casting.

In the above considerations of the pan and the handle the item that has so far not been discussed is the life of the items. The purchaser of the pan wants it to last, without problems, for a reasonable period of time. This is likely to be years. The handle should not break during this time or discolour or deteriorate when used and washed a large number of times. The pan should not wear thin or change its mechanical properties with frequent heating, exposure to hot liquids and washing-up liquids.

6.1.1 Stages in the selection process

As the above examples indicate, there are a number of stages involved in arriving at possible materials and processing requirements for a product. These can be summarised, in very simple terms, as follows:

1 Define the functions required of the product.
2 Consider a tentative design, taking into account any codes of practice, national or international standards.
3 Define the properties required of the materials.
4 Identify possible materials, taking into account availability in the required forms.
5 Identify possible processes which would enable the design to be realized.
6 Consider the possible materials and possible processes and arrive at a proposal for both. If not feasible, consider the design again and go back through the cycle.
7 Consider how the product will behave during its service life.

6.2 Costs

The total cost to the consumer of a manufactured article in service, i.e. the so-called *total life cost*, is made up of a number of items. These are:

1 The purchase price. This includes the costs of production, the fixed costs arising from factory overheads, administration, etc. and the manufacturer's profit. The costs of production include the cost of the materials and the cost of manufacture.
2 The cost of ownership. This includes such costs as those associated with maintenance, repair and replacement.

6.3 Failure in service

Failures in service can arise from:

1 Errors in the original design, e.g. the wrong material having been chosen.
2 The material used is in some way defective, e.g. the specification of the material to be obtained from the suppliers was not tight enough or was below specification and not detected by inspection.
3 Defects are introduced during the manufacturing process, e.g. heat treatment gives cracks as a result of quenching or perhaps incorrect assembly leading to misalignment and high stresses.
4 Deterioration in service, e.g. the product is exposed to unexpected corrosive environments or perhaps a temperature which results in changes in the microstructure of the material or perhaps poor maintenance which leads to nicks and gouges which act as stress raisers and a greater chance of failure due to fatigue.

6.3.1 The causes of failure

Failure can arise from a number of causes, e.g.

1 The stress level is just too high and the material yields and then breaks (Figure 6.1(a)). The material may show a ductile or a brittle form of fracture. With a ductile fracture there will have been quite significant yielding before the fracture occurs, with brittle fracture virtually none.

2 The material is subject to an alternating stress which results in fatigue failure. If you take a stiff piece of metal or plastic, and want to break it, then you will more likely flex the strip back and forth repeatedly, i.e. subject it to an alternating stress going from tension to compression to tension to compression and so on (Figure 6.1(b)). This is generally an easier way of causing the material to fail than applying a direct pull. The chance of fatigue failure occurring is increased, the greater the

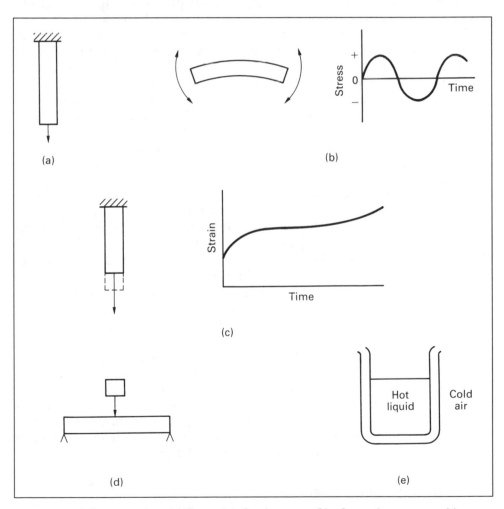

Figure 6.1 *Some modes of failure. (a) Static stress; (b) alternating stresses; (c) creep; (d) impact loading; (e) differential expansion*

amplitude of the alternating stresses. Stress concentrations produced by holes, surface defects and scratches, sharp corners, sudden changes in section, etc. can all help to raise the amplitude of the stresses at a particular point in the material and reduce its fatigue resistance.

3   The material is subject to a load which initially does not cause failure but the material gradually extends over a period of time until it fails. This is known as creep (Figure 6.1(c)). For most metals creep is negligible at room temperature but can become pronounced at high temperatures. For plastics, creep is often quite significant at ordinary temperatures and even more pronounced at higher ones. The creep of a metal, or a polymer, is determined by its composition and the temperature. For example, aluminium alloys will creep and fail at quite low stresses when the temperature rises above 200°C, while titanium alloys can be used at much higher temperatures before such a failure occurs.

4   A suddenly applied load (Figure 6.1(d)) causes failure, i.e. the energy at impact is greater than the impact strength of the material. Brittle materials have lower impact strengths than ductile ones.

5   The temperature changes and causes the properties to change in such a way that failure results, e.g. the temperature of a steel may drop to a level at which the material turns from being ductile to brittle and then easily fails as a result of perhaps an impact load.

6   A temperature gradient is produced and causes part of the product to expand more than another part, the resulting stresses resulting in failure, e.g. if you pour hot water into a cold glass then the inside of the glass tries to expand while the outside does not (glass has a low thermal conductivity), the result being that the glass cracks (Figure 6.1(e)).

7   The product is made of materials with differing coefficients of expansion. Thus when the temperature rises the different parts expand by different amounts, with the result that internal stresses are set up. These can result in distortion and possible failure.

8   Thermal cycling in which the temperature of the product repeatedly fluctuates will result in cycles of thermal expansion and contraction. If the material is constrained in some way then internal stresses will be set up and, as a consequence, the thermal cycling will result in alternating stresses being applied to the material. Fatigue failure can result. This is sometimes referred to as thermal fatigue.

9   Degradation because of the environment in which the material is situated, e.g. corrosion of iron leading to a reduction in the cross-section of a product and hence the resulting increase in stress leading to failure. Corrosion prevention, e.g. painting iron, is a major way of avoiding this type of problem. Plastics may become brittle as a result of exposure to the ultraviolet radiation in sunlight. This effect can be reduced by incorporating stabilizers with the polymer.

6.3.2   Examination of failures

A ductile material is characterized by having a significant plastic region to its stress–strain graph. When a ductile material has a gradually increas-

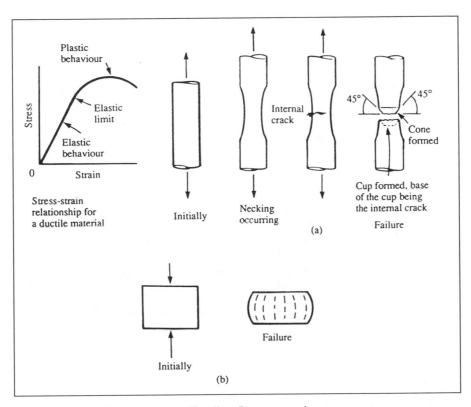

Figure 6.2 *Ductile failure. (a) Tensile; (b) compressive*

ing tensile stress applied then, when yielding starts, the cross-sectional area of the material becomes reduced, necking being said to occur (Figure 6.2). Eventually after a considerable reduction in cross-sectional area the material fails. The resulting fracture surfaces show a cone and cup formation. This occurs because, under the action of the increasing stress, small internal cracks form which gradually grow in size until there is an internal, almost horizontal, crack. The final fracture occurs when the material shears at an angle of 45° to the axis of the applied stress. Such a type of failure is referred to as a *ductile fracture*. Materials can also fail in a ductile manner in compression, such fractures resulting in a characteristic bulge and series of axial cracks around the edge of the material.

Another type of failure is known as *brittle fracture*. A brittle material has virtually no plastic region to its stress–strain graph. Thus when a brittle material fractures there is virtually no plastic deformation. Figure 6.3 shows possible forms of fracture in such a situation. The surfaces of the fractured material appear bright and granular due to the reflection of light from individual crystals.

Fatigue failure often starts at some point of stress concentration. This point of origin of the failure can be seen on the failed material as a smooth, flat, semicircular or elliptical region, often referred to as the nucleus.

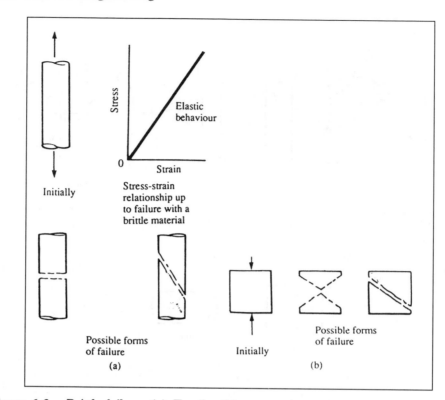

Figure 6.3 *Brittle failure. (a) Tensile; (b) compressive*

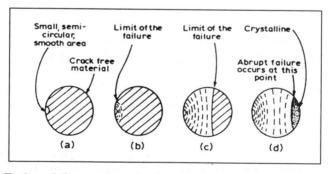

Figure 6.4 *Fatigue failure with a metal*

Surrounding the nucleus is a burnished zone with ribbed markings. This smooth zone is produced by the crack propagating relatively slowly through the material and the resulting fractured surfaces rubbing together during the alternating stressing of the component. When the component has become so weakened by the steadily spreading crack that it can no longer carry the load, the final abrupt fracture occurs. This region of abrupt failure has a crystalline appearance. Figure 6.4 shows the various stages in the growth of a fatigue failure.

6.4   Selection of materials

The following are some case studies, additional to that of the kitchen pan discussed in Section 6.1, in the selection of materials. They are designed to illustrate the processes involved in selection, and there are no doubt alternative solutions to these suggested. In addition, there are likely to be many more factors involved before a selection is made.

6.4.1   Car bodywork

The functions required of car bodywork are that it protects the engine and car occupants from the weather and provides a pleasing appearance. The requirements for the material are that it can be formed to the shapes required, it has a smooth and shiny surface, corrosion is not too significant, in service it is sufficiently tough to withstand small knocks, it is stiff, and is cheap and can be mass produced.

The shapes required and the fact that sheet material is required, together with the need for mass production, would suggest forming from sheet as the manufacturing process. Hot forming does present the problem of an unacceptable surface finish and so a material has to be chosen which allows for cold forming. This means a highly ductile material. Possibilities would be low-carbon steels or aluminium alloys. Table 6.1 shows some possibilities, giving the percentage elongations in the annealed state. On the basis of that information both the steels and the aluminium alloys could be used. In addition, both are reasonably tough and when given a coat of paint are reasonably resistant to corrosion and so can be expected to have a reasonable life.

The stress–strain graph of a low-carbon steel differs in form from that of an aluminium alloy in that the aluminium alloy shows a smooth transition from elastic to plastic deformation while the carbon steel shows some irregularities (Figure 6.5). The effect of this on the cold forming of carbon steel is to give some surface markings which do not occur with the aluminium alloy. The aluminium alloy thus has an advantage over the steel in giving a smoother surface when formed. Aluminium alloys also have another advantage – they have lower densities and so could lead to lower-weight cars. The carbon steel does, however, have some advantages, it work hardens more than the aluminium alloy (see Figure 5.9) and so gives a harder material. In addition, the steel has a higher tensile modulus than

Table 6.1   *Ductilities of carbon steel and aluminium alloys*

Material	% elongation
0.1% carbon steel	42
0.2% carbon steel	37
0.3% carbon steel	32
1.25% manganese–aluminium alloy	30
2.24% manganese–aluminium alloy	22

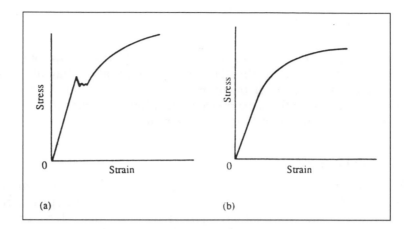

Figure 6.5   *Stress–strain graphs for (a) carbon steel; (b) aluminium alloy*

the aluminium and so is stiffer. The great advantage, outweighing all other considerations, is that carbon steel is much cheaper than aluminium alloy. Typically it is about half the price. Thus the choice is a low-carbon steel. In practice, the material used has less than 1% carbon.

6.4.2   Tennis rackets

The function of a tennis racket is to transmit power from the arm of the player to a tennis ball. The requirements for the frame and handle of a racket are high strength, high stiffness, low weight, toughness and ability to withstand impact loading, durability and not creeping or warping as a result of exposure to temperature or humidity changes, and ability to be processed into the required shape. Another requirement which requires a little explanation is the ability to damp out vibrations. When the ball hits the strings, vibrations occur. These are then transmitted through the frame of the racket to the arm of the player. If these vibrations are not reduced in amplitude in this transmission then the elbow of the player can suffer some damage, known as tennis elbow. The elbow joint does not like being vibrated. Cost will be a factor when considering tennis rackets for the general population but is less a requirement for rackets for professional tennis players.

The requirement for high strength and low weight can be translated into a need for a high value of strength/density, i.e. specific strength. Similarly the requirement for high stiffness and low weight results in a need for a high value of modulus/density, i.e. specific modulus. Possibilities would seem to be wood, metals and composites. Table 6.2 shows values for some possible materials.

Wood has the advantages that it is tough, has good specific strength, good damping properties for vibrations and is cheap. The specific stiffness could be better. Warping could be a problem. However, this can be overcome by using laminated wood, i.e. several pieces of wood with their fibres in different directions bonded together to give a laminate. This

Table 6.2   *Materials for tennis rackets*

Material	Specific strength (MPa/Mg m^{-3})	Specific stiffness (GPa/Mg m^{-3})	Relative toughness	Relative vibration damping	Relative cost
Wood: ash	107	20	Good	Good	Low
Wood: hickory	105	21	Good	Good	Low
Al–Cu alloy 2014 prec. hard.	15	25	Good	Poor	Medium
Al–Mg alloy 5050 annealed	54	25	Good	Poor	Medium
Mn steel 120M36 Q & T	90	27	Good	Poor	Medium
Ni–Cr–Mo steel 817M40 Q & T	115	27	Good	Poor	Medium
Composite: epoxy +60% carbon	890	90	Medium	Medium	High
Composite: epoxy +70% glass	750	25	Medium	Medium	High

combining together of pieces of wood also gives a method by which the shape of the racket can be obtained.

Aluminium alloy has the advantages of toughness and good specific stiffness. It is more expensive than wood. A problem, however, is that it has very poor vibration damping. Aluminium can be protected against corrosion attack by damp environments by anodizing. An aluminium racket can be made by bending extruded hollow sections into the required shape.

Steels can give high specific strengths and high specific stiffness. The steels with these high strengths are likely to be comparable in price with the aluminium alloys. Problems are, however, the very poor vibration damping and the poor corrosion resistance in a damp environment. A steel racket can be made by bending extruded hollow sections into the required shape.

Composite materials can be made which have the advantages of very high specific strengths, very high specific stiffnesses, reasonable vibration damping and tolerable toughness. The major problem, however, is the high cost of such materials. A composite racket can be made by injection moulding a melt of a polymer containing carbon fibres into a racket-shaped mould. This would give a racket with a solid composite for the frame and handle. The procedure that can then be adopted to improve the properties is, while the racket is still in the mould and only the outer skin of the composite has solidified, to pour out the liquid core so that when the racket solidifies there is a hollow tube. The tube can then be filled with a polyurethane foam. This improves the vibration damping of the racket.

In comparing the above, the composite material racket gives the best properties but is considerably more expensive than the others. It thus is more likely to be used by the professional tennis player. For cheapness and properties, wood is probably the next best material, followed by aluminium alloys with steel being the worst.

6.4.3 Small components for toys

Consider small components such as the wheels for, say, a small model toy car for use by a child. The functions required of the wheels are that they are safe and rotate on their axles. The materials thus need to be non-toxic, reasonably tough, not easily deformed by knocks, not brittle and cheap. Before considering possible materials, there is a BS 5665 which should be consulted. This specifies, in Part 1, material, construction and design requirements for toys, methods of test for certain properties, and requirements for packaging and marketing. Part 2 specifies categories of flammable materials not to be used in the manufacture of toys. Part 3 gives the requirements and methods of test for migration of antimony, arsenic, barium, cadmium, chromium, lead, mercury and selenium from toy materials.

The products are required to be cheap when produced in relatively large quantities and the products themselves are rather small. In the case of metals the obvious process is die casting. Though the initial cost of the die is high, a large number of components can be produced from one die and so the cost per component becomes relatively low. In the case of polymers the obvious choice is injection moulding. This also has a high die cost but large numbers of components can be produced from one die and hence the cost per component can be low. Both processes give good surface finish and dimensional accuracy.

In the case of metals, die casting limits the choice to those with relatively low melting points, i.e. aluminium, magnesium, zinc, lead and tin alloys. Table 6.3 shows relevant properties of these materials. Safety considerations (see the British Standard) rule out lead. Aluminium, magnesium and zinc are comparable in cost, with zinc tending to have the lower cost per unit weight. Tin is more expensive than these alloys. Zinc has a lower melting point than aluminium or magnesium and in the as-cast condition has the highest tensile strength. Thus zinc would seem to be the best metal choice for the product.

Zinc is very widely used for die casting. Its low melting point and excellent fluidity make it one of the easiest metals to cast. Small parts of complex shape and thin wall sections can be produced. Zinc alloys have relatively good mechanical properties and can be electroplated.

Polymers are a possible alternative to metals. Since the forming method is to be injection moulding, the materials are restricted to thermoplastics.

Table 6.3 Die casting alloys in the as-cast condition at 20°C

Alloy	Density (Mg m^{-3})	Melting point (°C)	Strength (MPa)
Aluminium	2.7	600	150
Lead	11.3	320	20
Magnesium	1.8	520	150
Tin	7.3	230	12
Zinc	6.7	380	280

Table 6.4 *Comparision of zinc and thermoplastics at 20°C*

Material	Strength (MPa)	Modulus (GPa)	Density (Mg m⁻³)	Relative cost (m³)
ABS	50	2.3	1.02-1.07	1
Nylon 6	60	3.2	1.13-1.14	2
Polycarbonate	65	2.3	1.2	2
Zinc alloy	280	103	6.7	3

The choice is then of polymers which are relatively stiff. Table 6.4 shows how the properties of possible polymers compare with those of zinc. The mechanical properties of the zinc alloy are superior to those of thermoplastics, it having higher strength, higher tensile modulus, and being tougher and more resistant to fatigue and creep. Where light weight is required then polymers have the advantage, having densities of the order of one-sixth that of zinc. Where coloured surfaces are required then polymers have the advantage since pigments can be incorporated into the polymer mix. However, if electroplating is required then zinc has the advantage. On cost per unit weight then zinc is cheaper, but the interest is likely to be cost per unit volume and on this basis polymers are likely to be cheaper.

On the basis of the above considerations, it is likely that zinc would be used where fine detail and colour of surface is not a requirement but a metallic surface as a result of electroplating is required. Otherwise polymers, possibly ABS, would be used.

Problems

1 Make reasoned proposals for materials and processes for the following products:
   (a) domestic window catches,
   (b) structural I-beams for use in building construction,
   (c) rainwater gutters and drainpipes,
   (d) a domestic washing-up bowl,
   (e) pipe through which sea water can be pumped,
   (f) small fan in a vacuum cleaner,
   (g) the lenses for the rear lights of cars,
   (h) a camshaft for a car,
   (i) casing for a hand-held power tool,
   (j) the blades for a hover mower.
2 Investigate the materials used with the following products and give reasons why they might have been chosen in preference to others and the processes that might have been used:
   (a) the casing for mains electric plugs,
   (b) spades,
   (c) domestic cold and hot water pipes,

(d) the casing for the body of a vacuum cleaner,

(e) joists to support floors in a small house.

3   For the following consult the British Standard specified and then present reasons for the choice of materials and processes for the specified products:

(a) BS 4654 Hooks for lifting freight containers,

(b) BS 3388 Forks, shovels and spades,

(c) BS 3441 Tanks for the transport of milk,

(d) BS 3531 Section 5.2 Screwdrivers,

(e) BS 4109 Copper wire for electrical purposes,

(f) BS 4344 Pulley blocks for use with natural and synthetic fibre ropes,

(g) BS 4637 Coiled springs,

(h) BS 6728 Hot water bottles made from PVC compounds,

(i) BS 7413/4 Unplasticized PVC for window frames.

7 Safety parameters

Outcomes At the end of this chapter you should be able to:

- Recognize the responsibility of the individual to others and the organization in the need to work within required safety parameters, i.e. Health and Safety at Work.
- Assess the risks in the workplace associated with the use, handling, processing, storage and disposal of materials, i.e. with the Control of Substances Hazardous to Health.

Note: this chapter gives only a brief overview of health and safety at work. For a detailed consideration the reader is referred to the many publications of the Health and Safety Executive, a booklet is available from them listing the publications available.

7.1 Health and safety at work

Safety in a company is everyone's responsibility. While there may be some employees, such as safety officers or safety representatives, who have special responsibilities, all the employees and the employers have responsibilities. These responsibilities are a legal requirement under the Health and Safety at Work Act 1974.

The *Health and Safety at Work Act 1974* provides a comprehensive and integrated system of law in relation to the health, safety and welfare of people at work and for the protection of the general public against risks to health and safety arising out of, or in connection with, the activities of persons at work. The Act consists of four parts, Part I being concerned with health, safety and welfare in connection with work and the control of dangerous substances and certain emissions into the atmosphere. Part II deals with the Employment Medical Advisory Service. Part III covers building regulations. Part IV deals with miscellaneous and general matters. The Act did not immediately repeal earlier legislation and regulations but allowed them to remain current until revoked and replaced by new regulations or Codes of Practice issued under the act.

Part I is of particular relevance in connection with this book and has four basic objectives:

1 To secure the health, safety and welfare of people at work.
2 To protect persons other than persons at work against risks to health or safety arising out of or in connection with the activities of those at work.
3 To control the keeping and use of explosive or highly flammable or otherwise dangerous substances and generally prevent the unlawful acquisition, possession and use of such substances.
4 To control the emission into the atmosphere of noxious or offensive substances.

7.1.1 Employers' responsibilities

Under Part I of the Act, general duties are laid down for employers. These are to ensure, as far as is reasonably practicable, the health, safety and welfare at work of all employees. This extends to:

1 *Plant and equipment*
   The provision and maintenance of plant and systems of work that are, so far as is reasonably practicable, safe and without risk to health.
2 *Handling, storage and transport*
   Arrangements for ensuring, so far as is reasonably practicable, safety and absence of risks to health in connection with the use, handling, storage and transport of articles and substances.
3 *Information, training and supervision*
   The provision of such information, instruction, training and supervision as is necessary to ensure, so far as is reasonably practicable, the health and safety at work of employees. As a consequence, this requires employers to identify potential hazards.
4 *Safe premises*
   So far as is reasonably practicable, maintaining the place of work in a condition that is safe and without risks to health and the provision and maintenance of means of access and egress from it that are safe and without such risks.
5 *Safe working environment*
   The provision and maintenance of a working environment for employees that is, so far as is reasonably practicable, without risks to health and adequate as regards facilities and arrangements for employee welfare at work.

In addition, the employer must:

1 Provide a written statement of the organization's general policy with respect to the health and safety of the employees and how it will be carried out.
2 Safety representatives, elected from and by the employees, must be consulted by the employer on safety matters.

3   A safety committee must be established if the safety representatives so request.

In *Essentials of Health and Safety at Work*, published by the Health and Safety Executive (1992), eleven steps are listed for employers to review the risks and safeguards in their operations and where they might benefit from a safety-improvement plan. These steps are:

1   Know your legal duties.
2   Provide safe methods by finding out about safe working methods for the industry and making certain that everyone is aware of them.
3   Organize the duties of those who are to have specific responsibilities for safety.
4   Prepare a safety policy (see the Health and Safety Executive leaflet (6): *Writing a safety policy statement: advice to employers*).
5   Train the staff
6   Train yourself to identify hazards and ways of delaying them.
7   Check your performance, i.e. check that rules are being followed.
8   Organize your information by keeping safety documentation separately filed.
9   Investigate when things go wrong.
10 Prepare your safety-improvement plan if changes are required.
11 Maintain interest in safety by seeking the cooperation and active commitment of supervisors, employees and safety representatives.

7.1.2   Employees' responsibilities

Under the Act, employees also have duties with respect to health and safety at work. While at work, employees should:

1   Take reasonable care of their own health and safety and of other persons who may be affected by his or her acts of omissions at work.
2   To cooperate with the employer as regards any duty or requirement imposed on the employer or any other person as a consequence of statutory provisions in order to enable the duty or requirement to be complied with.

In addition, no person shall intentionally or recklessly interfere with or misuse anything provided in the interests of health, safety or welfare as a consequence of statutory provisions. For example, an employee removing the safety guards from a machine could be prosecuted under the Act.

7.1.3   Health and Safety Commission and Executive

Two bodies were set up by the Act: the Health and Safety Commission and the Health and Safety Executive. The Commission is charged with the task of making arrangements for carrying out research, the provision of training and information, advisory services and the development of regulations. The Executive exercises, on behalf of the Commission, such of its functions as the Commission directs it to exercise. The tasks of enforcing the statutory provisions is given to the Executive, in cooperation with local authorities and other enforcement bodies.

Every enforcing authority may appoint inspectors. Such inspectors may visit workplaces without notice. They may want to examine the safety, health and welfare situation in the company, or perhaps investigate an accident or complaint. They have the right to talk to employees and safety representatives, take such samples and photographs as they consider relevant and in some situations impound dangerous equipment. If, as a result of such investigations an inspector considers a situation to be contravening relevant statutory provisions, he or she may serve an improvement notice requiring the situation to be remedied within a specified time. If there is a risk of personal injury, the inspector can issue a prohibition notice. Such a notice requires the activity concerned to cease until the situation has been remedied. Failure to comply with notices can result in prosecution and fines and/or imprisonment.

There are a number of publications available from Health and Safety Executive offices. These include leaflets such as *HSW Act: the Act outlines* (2), *HSW Act: advice to employers* (3), *HSW Act: advice to employees* (5), and regulations, approved codes of practice and guidance literature (7). In addition there is *Guide to the HSW Act* (ISBN 0 11 885555 7).

7.2 Safe work systems

Employers have the duty that they should provide systems of work that are, so far as is reasonably practicable, safe and without risks to health. This requires the establishment of safe work systems and a consideration of such factors as:

1 *Health*
   This, for example, means considering the hazards to health of exposure to certain chemicals, of what happens when chemicals are spilt, etc.
2 *Safety*
   This means, for example, a consideration of the hazards that can occur when a machine or its guard fails, or an operator chooses to do a job in a different way, etc.
3 *Permits to work*
   The issuing of a safe written procedure may be adequate for some jobs, but where the risks are very high there can be the need to initiate a formal permit system. This permit states exactly what work is to be done and when and whether the work is assessed and checked by a responsible person. Those doing the job sign the permit to show that the hazards and precautions necessary are understood.
4 *The needs of individuals*
   This may include protective clothing and equipment, seating and working space being considered in relation to the needs of each individual. There are also such considerations as whether they are able to understand safety instructions and can work safely if under medication, have some handicap, etc.

5   *Maintenance*
In order to ensure continuing safety, equipment, buildings and plant need proper maintenance.

6   *Monitoring*
The system needs checking to see that rules and precautions are being followed and continue to deal with the risks. New hazards may be found to be introduced by changes in staff, materials, equipment, etc.

7.2.1   Protective clothing and equipment

Wherever possible, employers should eliminate or control risks so that protective clothing and equipment do not have to be used. Some jobs, however, require protective clothing or equipment specified by law. The following is an indication of some of the hazards and protection that might be used:

Eyes
Hazards: chemical or metal splashes, dust, projectiles, gas, radiation.

Protection: spectacles, goggles, face screens, helmets.

Head and neck
Hazards: falling objects, bumping head, hair entanglement, chemicals.

Protection: helmets, bump caps, hairnets, caps, skull-caps.

Hearing
Hazards: impact noise, high levels of sound, high- and low-frequency sound.

Protection: earplugs, muffs

Hands and arms
Hazards: abrasion, temperature extremes, cuts, chemicals, electric shock, skin infections, vibration.

Protection: gloves, gauntlets, wrist cuffs, armlets.

Feet and legs
Hazards: wet, slipping, cuts, falling objects, heavy pressures, metal and chemical splashes, abrasion.

Protection: safety boots, gaiters, leggings.

Respiratory protection
Hazards: toxic and harmful dusts, gases and vapours, micro-organisms.

Protection: disposable respirators, mask respirators, fresh air hose equipment.

The body
Hazards: heat, cold, weather, chemical or metal splashes, impact, contaminated dust, entanglement of clothing.

Protection: overalls, boiler suits, warehouse coats, donkey jackets, aprons, specialist protective clothing.

7.2.2 Accidents and emergencies

There is a need to plan for dealing with emergencies, whether they are simple or major incidents. People need to be told what might happen, how the alarm will be raised, what to do, where to go, who will be in control, and what essential actions need to be taken, e.g. closing down plant. They need training in emergency procedures. Access-ways need to be kept clear for emergency services and escape routes. Fire-fighting equipment, electrical isolators and shut-off valves need to be clearly labelled. Emergency equipment needs testing regularly.

After an accident or a serious incident, the immediate emergency should be dealt with, any injuries treated and the premises or plant made safe. Any injuries should be recorded in an accident book. If applicable, the incident should be reported to the inspector. The Reporting of Injuries, Diseases and Dangerous Occurrences Regulations 1985 require an employer to report immediately by telephone if as a result of or in connection with work someone dies, receives a major injury, is seriously affected by, for example, electric shock or poisoning, or there is a dangerous occurrence (for further information, see Health and Safety Executive leaflet (11) *Reporting an injury or a dangerous occurrence*). Also, a written report needs to be to sent within seven days confirming the telephone report and a notification of any injury which stops someone doing their normal job for more than three days, whether certain diseases are suffered by the workers, and certain types of events involving flammable gas in domestic or other premises (see the Health and Safety Regulations booklet (23) *Guide to reporting of injuries, diseases and dangerous occurrence regulations 1985*).

7.3   A safe and healthy environment

The employer is charged with the provision and maintenance of a working environment for the employees that is, as far as is practicable, safe, without risks to health and adequate as regards facilities and arrangement for their welfare at work. The employer has thus to be concerned with such facilities as hygiene and welfare; cleanliness; floors and gangways being kept clean, dry and not slippery; seats, machine controls, instruments and tools being designed for the best control, use and posture; the place of work being safe with adequate space for easy movement and safe machine adjustment, no tripping hazards, emergency provisions, etc.; lighting that gives good general illumination with no glare, adequate emergency lighting, etc.; and a comfortable environment with a suitable working temperature, good ventilation, acceptable noise levels, etc.

Health hazards can arise in a number of ways since people at work are exposed to a wide range of substances. The Control of Substances Hazardous to Health Regulations 1988 deal with such hazards.

7.3.1   Control of Substances Hazardous to Health

The *Control of Substances Hazardous to Health Regulations 1988* (COSHH) impose duties on employers to protect employees and other persons who may be exposed to substances hazardous to health and also certain duties on employees. The term 'substances hazardous to health' covers substances:

1 Listed in an approved list as dangerous and for which the general indication of the nature of risk is specified as very toxic, toxic, harmful, corrosive or irritant;
2 For which a maximum exposure limit has been specified or for which there is an occupational exposure standard;
3 Any microorganism hazardous to health;
4 Dust of any kind when present at a substantial concentration in air;
5 Any other substance which could create a hazard to health comparable to those listed above.

Employers are required to:

1 Assess the risk of exposure to hazardous substances and the precautions which should be taken to prevent exposure, or if this is not reasonably practicable, to achieve adequate control.
2 Introduce appropriate control measures to prevent or control the risk.
3 Ensure that the control measures are used.
4 Ensure that the control measures are properly maintained and periodically examined.
5 Where necessary, the exposure of workers should be monitored.
6 Where necessary, carry out health surveillance of workers.
7 Inform, instruct and train employees and their representatives of the risks involved, the precautions to be taken and the results of monitoring.

As an indication of the types of hazards needing to be controlled, consider the processing of plastics (for further information, see *The Application of COSHH to Plastics Processing*, Health and Safety Executive, 1990). Moulding and extrusion machines have automatic alarm systems to limit overheating of polymers. These may not always be correctly set. Heating thermoplastic materials to beyond their processing range causes them to decompose and produce fumes. The normal processing of polymers may also give rise to fumes. In some cases these fumes can be toxic. In addition, there may be dust hazards during the supply of dry materials to the feed units of the machines. The hazards posed by such fumes and dust needs to be assessed and guarded against.

Problems

1 What are the general duties imposed by the Health and Safety at Work Act 1974 on (a) employers and (b) employees?
2 Explain the functions of (a) improvement notices and (b) prohibition notices.

3   Obtain one of the following Health and Safety Executive booklets and produce a paper outlining the consequences of such guidance on operations.
    (a) *Safety devices for hand- and foot-operated presses (3)*,
    (b) *Safety in drop forging hammers (12)*,
    (c) *Safety in the stacking of materials (47)*.

4   Discuss the following situations and consider what actions should be taken or what consequences could occur:
    (a) A worker removes the safety guards from a machine because they reduce the number of items he can produce and hence his wages, which are based on piece-rate.
    (b) The company is short of storage space for a large consignment of goods that has just arrived and so they temporarily stack them in the passage ways leading to the main exit from the factory floor.
    (c) A worker is just about to become married and in celebration fellow-workers set off the fire extinguishers.
    (d) An inspector wants to go into the tool room but the management tell him that they will not allow it because the work there is commercially highly secret and he might tell their competitors.
    (e) An inspector wants to talk to a worker about his machine but the worker refuses to talk to him.
    (f) A worker is injured and as a result is off work for a month. The employers pay him or her during that time but take no other action.

5   What protective clothing and equipment might be suitable in the following situations?
    (a) Chemicals may splash into the eyes.
    (b) A press is very noisy during operation.
    (c) A storekeeper has to move sheets of metal.

Appendix 1: Units

The International System (SI) of units has seven base units, these being:

Length	metre	m
Mass	kilogram	kg
Time	second	s
Electric current	ampere	A
Temperature	kelvin	K
Luminous intensity	candela	cd
Amount of substance	mole	mol

In addition, there are two supplementary units: the radian and the steradian.

The SI units for other physical quantities are formed from the base units via the equation defining the quantity concerned. Thus, for example, volume is defined by the equation volume = length cubed, thus the unit of volume of the length unit cubed, i.e. m^3. Density is defined by the equation density = mass/volume and thus has the unit of mass divided by the unit of volume, i.e. kg/m^3. Velocity is defined by the equation velocity = change of displacement/time taken and thus has the unit of length divided by the unit of time, i.e. m/s. Acceleration is defined by the equation acceleration = change in velocity/time taken and thus has the unit of velocity divided by the unit of time, i.e. (m/s)/s or m/s^2.

Some of the derived units are given special names. Thus, for example, the unit of force is defined by the equation force = mass × acceleration and is kg m/s or kg m s^{-1}. This unit is given the name newton (N). Thus 1 N = 1 kg m s^{-1}. The unit of stress is given by the equation stress = force/area and has the derived unit of N/m^2. This unit is given the name pascal (Pa). Thus 1 Pa = 1 N/m^2.

Certain quantities are defined as the ratio of two comparable quantities. Thus, for example, strain is defined as change in length/length. It thus is expressed as a pure number with no units because the derived unit would be m/m.

Standard prefixes are used for multiples and submultiples of units, the SI preferred ones being multiples of 10^3. The following are the standard prefixes:

Multiplication factor		Prefix	
1 000 000 000 000 000 000 000 000 = 10^{24}	yotta	Y
1 000 000 000 000 000 000 000 = 10^{21}	zetta	Z
1 000 000 000 000 000 000 = 10^{18}	exa	E
1 000 000 000 000 000 = 10^{15}	peta	P
1 000 000 000 000 = 10^{12}	tera	T
1 000 000 000 = 10^{9}	giga	G
1 000 000 = 10^{6}	mega	M
1 000 = 10^{3}	kilo	k
100 = 10^{2}	hecto	h
10 = 10	deca	da
0.1 = 10^{-1}	deci	d
0.01 = 10^{-2}	centi	c
0.001 = 10^{-3}	milli	m
0.000 001 = 10^{-6}	micro	μ
0.000 000 001 = 10^{-9}	nano	n
0.000 000 000 001 = 10^{-12}	pico	p
0.000 000 000 000 001 = 10^{-15}	femto	f
0.000 000 000 000 000 001 = 10^{-18}	atto	a
0.000 000 000 000 000 000 001 = 10^{-21}	zepto	z
0.000 000 000 000 000 000 000 001 = 10^{-24}	yocto	y

Thus, for example, 1000 N can be written as 1 kN, 1 000 000 Pa as 1 MPa, 1 000 000 000 Pa as 1 GPa, 0.001 m as 1 mm, and 0.000 001 A as 1 μA. Note that often the unit N/mm^2 is used for stress, 1 N/mm^2 is 1 MPa.

For further information about SI units, including their definitions, the reader is referred to *SI The International System of Units*, edited by R. J. Bell (National Physical Laboratory, HMSO). This is the approved translation of the International Bureau of Weights and Measures publication.

Other units which the reader may come across are fps (foot-pound-second) units which still are often used in the USA. On that system the unit of length is the foot (ft), with 1 ft = 0.3048 m. The unit of mass is the pound (lb), with 1 lb = 0.4536 kg. The unit of time is the second, the same as the SI system. With this system the derived unit of force, which is given a special name, is the poundal (pdl), with 1 pdl = 0.1383 N. However, a more common unit of force is the pound force (lbf). This is the gravitational force acting on a mass of 1 lb and consequently, since the standard value of the acceleration due to gravity is 32.174 ft/s^2, then 1 lbf = 32.174 pdl = 4.448 N. The similar unit the kilogram force (kgf) is sometimes used. This is the gravitational force acting on a mass of 1 kg and consequently, since the standard value of the acceleration due to gravity is 9.806 65 m/s^2, then 1 kgf = 9.806 65 N. A unit often used for stress in the USA is the psi, or pound force per square inch. 1 psi = 6.895 Pa.

Appendix 2: Terminology

The following are some of the terms commonly encountered in discussing materials for engineering. For terms for materials see Appendix 3.

Additives   Plastics and rubbers almost invariably contain, in addition to the polymer or polymers, other materials, i.e. additives. These are added to modify the properties and cost of the material.

Ageing   This term is used to describe a change in properties that occurs with certain metals due to precipitation occurring, there being no change in chemical composition.

Alloy   This is a metal which is a mixture of two or more elements.

Amorphous   An amorphous material is a non-crystalline material, i.e. it has a structure which is not orderly.

Annealing   This involves heating to and holding at a temperature which is high enough for recrystallization to occur and which results in a softened state for a material after a suitable rate of cooling, generally slowly. The purpose of annealing can be to facilitate cold working, improve machine-ability and mechanical properties, etc.

Anodizing   This term is used to describe the process, generally with aluminium, whereby a protective coating is produced on the surface of the metal by converting it to an oxide.

Austenite   This term describes the structure of a face-centred cubic iron crystalline structure which has carbon atoms in the gaps in the face-centred iron.

Bend, angle of   The results of a bend test on a material are specified in terms of the angle through which the material can be bent without breaking. The greater the angle, the more ductile the material.

Brinell number   The Brinell number is the number given to a material as a result of a Brinell test and is a measure of the hardness of a material. The larger the number, the harder the material.

Brittle failure   With brittle failure a crack is initiated and propagates prior to any significant plastic deformation. The fracture surface of a metal with a brittle fracture is bright and granular due to the reflection of light from individual crystal surfaces. With polymeric materials the fracture surface may be smooth and glassy or somewhat splintered and irregular.

129

Brittle material   A brittle material shows little plastic deformation before fracture. The material used for a china teacup is brittle. Thus because there is little plastic deformation before breaking, a broken teacup can be stuck back together again to give the cup the same size and shape as the original.

Carburizing   This is a treatment which results in a hard surface layer being produced with ferrous alloys. The treatment involves heating the alloy in a carbon-rich atmosphere so that carbon diffuses into the surface layers, then quenching to convert the surface layers to martensite.

Case hardening   This term is used to describe processes in which, by changing the composition of surface layers of ferrous alloys, a hardened surface layer can be produced.

Casting   This is a manufacturing process which involves pouring liquid metal into a mould or, in the case of plastics, the mixing of the constituents in a mould.

Cementite   This is a compound formed of iron and carbon, often referred to as iron carbide. It is a hard and brittle material.

Charpy test value   The Charpy test is used to determine the response of a material to a high rate of loading and involves a test piece being struck a sudden blow. The results are expressed in terms of the amount of energy absorbed by the test piece when it breaks. The higher the test value, the more ductile the material.

Cold working   This is when a metal is subject to working at a temperature below its recrystallization temperature.

Composite   This is a material composed of two different materials bonded together in such a way that one serves as the matrix surrounding fibres or particles of the other.

Compressive strength   The compressive strength is the maximum compressive stress a material can withstand before fracture.

Copolymer   This is a polymeric material produced by combining two or more monomers into a single polymer chain.

Corrosion resistance   This is the ability of a material to resist deterioration by reacting with its immediate environment. There are many forms of corrosion and so there is no unique way of specifying the corrosion resistance of a material.

Creep   Creep is the continuing deformation of a material with the passage of time when it is subject to a constant stress. For a particular material the creep behaviour depends on both the temperature and the initial stress, the behaviour also depending on the material concerned.

Crystalline   This term is used to describe a structure in which there is a regular, orderly arrangement of atoms or molecules.

Damping capacity   The damping capacity is an indicator of the ability of a material to suppress vibrations.

Density   Density is mass per unit volume.

Dielectric strength   The dielectric strength is a measure of the highest potential difference an insulating material can withstand without electric breakdown. It is the breakdown voltage divided by the thickness of the material.

Ductile failure   With ductile failure there is a considerable amount of plastic deformation prior to failure. With metals the fracture shows a typical cone and cup formation and the fracture surfaces are rough and fibrous in appearance.

Ductile materials   Ductile materials show a considerable amount of plastic deformation before breaking. Such materials have a large value of percentage elongation.

Elastic limit   The elastic limit is the maximum force or stress at which on its removal the material returns to its original dimensions.

Electrical conductance   This is the reciprocal of the electrical resistance and has the unit of the siemen (S). It is thus the current through a material divided by the voltage across it.

Electrical conductivity   The electrical conductivity is defined by

$$\text{Conductivity} = \frac{L}{RA}$$

where R is the resistance of a strip of the material of length L and cross-sectional area A. Conductivity has the unit of S/m. The IACS specification of conductivity is based on 100% corresponding to the conductivity of annealed copper at 20°C, and all other materials are then expressed as a percentage of this value.

Electrical resistance   This is the voltage across a material divided by the current through it, the unit being the ohm (Ω).

Electrical resistivity   The electrical resistivity is defined by

$$\text{Resistivity} = \frac{RA}{L}$$

Resistivity has the unit Ω m.

Expansion, coefficient of linear   The coefficient of linear expansion is a measure of the amount by which a unit length of a material will expand when the temperature rises by one degree. It is defined by

$$\text{Coefficient} = \frac{\text{change in length}}{\text{length} \times \text{temp. change}}$$

It has the unit °C^{-1} or K^{-1}.

Expansivity, linear   This is an alternative name for the coefficient of linear expansion.

Fatigue life   The fatigue life is the number of stress cycles to cause failure.

Ferrite   This term is usually used for a structure consisting of carbon atoms lodged in body-centred cubic iron. Ferrite is comparatively soft and ductile.

Fracture toughness   The plane strain fracture toughness is an indicator of whether a crack will grow or not and thus is a measure of the toughness of a material when there is a crack present.

Full hard   This term is used to describe the temper of alloys. It corresponds to the cold-worked condition beyond which the material can no longer be worked.

Grain   This term is used for a crystalline region within a metal, i.e. a region of orderly packed atoms.

Half-hard   This term is used to describe the temper of alloys. It corresponds to the cold-worked condition half-way between soft and full hard.

Hardness   The hardness of a material may be specified in terms of some standard test involving indentation, e.g. the Brinell, Vickers and Rockwell tests, or scratching of the surface of the material, the Moh test.

Heat treatment   This term is used to describe the controlled heating and cooling of metals in the sold state for the purpose of altering their properties.

Hooke's law   When a material obeys Hooke's law its extension is directly proportional to the applied stretching forces.

Hot working   This is when a metal is subject to working at a temperature in excess of its recrystallization temperature.

Impact properties   See *Charpy test value* and *Izod test values*.

Izod test value   The Izod test is used to determine the response of a material to a high rate of loading and involves a test piece being struck a sudden blow. The results are expressed in terms of the amount of energy absorbed by the test piece when it breaks. The higher the test value, the more ductile the material.

Limit of proportionality   Up to the limit of proportionality the extension is directly proportional to the applied stretching forces, i.e. the strain is proportional to the applied stress.

Malleability   This describes the ability of metals to permit plastic deformation in compression without rupturing.

Martensite   This is a general term used to describe a form of structure. In the case of ferrous alloys it is a structure produced when the rate of cooling from the austenitic state is too rapid to allow carbon atoms to diffuse out of the face-centred cubic form of austenite and produce the body-centred form of ferrite. The result is a highly strained hard structure.

Melting point   This is the temperature at which a material changes from solid to liquid.

Moh scale   This is a scale of hardness arrived at when considering the ease of scratching a material. It is a scale of 10, with the higher the number, the harder the material.

Monomer   This is the unit, or mer, consisting of a relatively few atoms which are joined together in large numbers to form a polymer.

Nitriding   This is a treatment in which nitrogen diffuses into surface layers of a ferrous alloy and hard nitrides are produced, hence a hard surface layer.

Orientation   A polymeric material is said to have an orientation, uniaxial or biaxial, if during the processing of the material the molecules become aligned in particular directions. The properties of the material in such directions are markedly different from those in other directions.

Pearlite   This is a lamellar structure of ferrite and cementite.

Percentage elongation   The percentage elongation is a measure of the ductility of a material, the higher the percentage, the greater the ductility. It is the change in length which has occurred during a tensile test to breaking expressed as a percentage of the original length.

$$\% \text{ elongation} = \frac{\text{final} - \text{initial length}}{\text{initial length}} \times 100$$

Percentage reduction in area   This is a measure of the ductility of a material and is the change in cross-sectional area which has occurred during a tensile test to breaking expressed as a percentage of the original cross-sectional area.

Precipitation hardening   This is a heat treatment process which results in a precipitate being produced in such a way that a harder material is produced.

Proof stress   The 0.2% proof stress is defined as that stress which results in a 0.2% offset, i.e. the stress given by a line drawn on the stress–strain graph parallel to the linear part of the graph and passing through the 0.2% strain value. The 0.1% proof stress is similarly defined. Proof stresses are quoted when a material has no well-defined yield point.

Quenching   This is the method used to produce rapid cooling. In the case of ferrous alloys it involves cooling from the austenitic state by immersion in cold water or an oil bath.

Recovery   This term is used for the treatment involving the heating of a metal so as to reduce internal stresses.

Recrystallization   This is generally used to describe the process whereby a new, strain-free grain structure is produced from that existing in a cold-worked metal by heating.

Resilience   This term is used with elastomers to give a measure of the 'elasticity' of a material. A high-resilience material will suffer elastic collisions when a high percentage of the kinetic energy before the collision is returned to the object after it. A less resilient material would lose more kinetic energy in the collision.

Rockwell test value   The Rockwell test is used to give a value for the hardness of a material. There are a number of Rockwell scales and thus the scale being used must be quoted with all test results.

Ruling section   The limiting ruling section is the maximum diameter of round bar at the centre of which the specified properties may be obtained.

Secant modulus   For many polymeric materials there is no linear part of the stress–strain graph and thus a tensile modulus cannot be quoted. In such cases the secant modulus is used. It is the stress at a value of 0.2% strain divided by that strain.

Shear   When a material is loaded in such a way that one layer of the material is made to slide over an adjacent layer then the material is said to be in shear.

Shear strength   The shear strength is the shear stress required to produce fracture.

Sintering   This is the process by which powders are bonded by molecular or atomic attraction as a result of heating to a temperature below the melting points of the constituent powders.

Solution treatment   This heat treatment involves heating an alloy to a suitable temperature, holding at that temperature long enough for one or more constituent elements to enter into the crystalline structure, and then cooling rapidly enough for these to remain in solid solution.

Specific gravity   The specific gravity of a material is the ratio of its density compared with that of water.

Specific heat capacity   The amount by which the temperature rises for a material when there is a heat input depends on its specific heat capacity. The higher the specific heat capacity, the smaller the rise in temperature per unit mass for a given heat input.

$$\text{Specific heat capacity} = \frac{\text{heat input}}{\text{mass} \times \text{temp. change}}$$

Specific heat capacity has the unit $J\ kg^{-1}\ K^{-1}$.

Specific stiffness   This is the modulus of elasticity divided by the density.

Specific strength   This is the strength divided by the density.

Stiffness   The property is described by the modulus of elasticity.

Strain   The engineering strain is defined as the ratio (change in length)/(original length) when a material is subject to tensile or compressive forces. Shear strain is the ratio (amount by which one layer slides over another)/(separation of the layers). Because it is a ratio, strain has no units, though it is often expressed as a percentage. Shear strain is usually quoted as an angle in radians.

Strength   See *Compressive strength*, *Shear strength* and *Tensile strength*.

Stress   In engineering, tensile and compressive stress is usually defined as (force)/(initial cross-sectional area). The true stress is (force)/(cross-sectional area at that force). Shear stress is the (shear force)/(area resisting shear). Stress has the unit Pa (pascal) with $1\ Pa = 1\ N\ m^{-2}$.

Stress relieving   This is a treatment to reduce residual stresses by heating the material to a suitable temperature, followed by slow cooling.

Stress–strain graph   The stress–strain graph is usually drawn using the engineering stress (see *Stress*) and engineering strain (see *Strain*).

Surface hardening   This is a general term used to describe a range of processes by which the surface of a ferrous alloy is made harder than its core.

Temper   This term is used with non-ferrous alloys as an indication of the degree of hardness/strength, with expressions such as hard, half-hard, three-quarters hard being used.

Tempering   This is the heating of a previously quenched material to produce an increase in ductility.

Tensile modulus   The tensile modulus, or Young's modulus, is the slope of the stress-strain graph over its initial straight-line region.

Tensile strength   This is defined as the maximum tensile stress a material can withstand before breaking.

Thermal conductivity   The rate at which energy is transmitted as heat through a material depends on a property called the thermal conductivity. The higher the thermal conductivity, the greater the rate at which heat is conducted. Thermal conductivity is defined by

$$\text{Thermal conductivity} = \frac{\text{rate of transfer of heat}}{\text{area} \times \text{temp. gradient}}$$

Thermal conductivity has the unit $W\ m^{-2}\ K^{-1}$.

Thermal expansivity   See *Expansion, coefficient of linear*.

Toughness   This property describes the ability of a material to absorb energy and deform plastically without fracturing. It is usually measured with the Izod test or the Charpy test. Another form of measure is the fracture toughness. See *Fracture toughness*.

Transition temperature   The transition temperature is the temperature at which a material changes from giving a ductile failure to giving a brittle failure.

Vickers test results   The Vickers test is used to give measure of hardness. The higher the Vickers hardness number, the greater the hardness.

Water absorption   This is the percentage gain in weight of a polymeric material after immersion in water for a specified amount of time under controlled conditions.

Wear resistance   This is a subjective comparison of the wear resistance of materials. There is no standard test.

Work hardening   This is the hardening of a material produced as a consequence of working subjecting it to plastic deformation at temperatures below those of recrystallization.

Yield point   For many metals, when the stretching forces applied to a test piece are steadily increased a point is reached when the extension is no longer proportional to the applied forces and the extension increases more rapidly than the force until a maximum force is reached. This is called the upper yield point. The force then drops to a value called the lower yield point before increasing again as the extension is continued.

Young's modulus   See *Tensile modulus*.

	# Appendix 3: Materials

This appendix lists commonly encountered engineering metals, polymers and ceramics and their main characteristics.

Engineering metals

The following is an alphabetical listing of metals, each being listed according to the main alloying element, with their key characteristics. It is not a comprehensive list of all metallic elements, just those commonly encountered in engineering.

Aluminium   Used in commercially pure form and alloyed with copper, manganese, silicon, magnesium, tin and zinc. Alloys exist in two groups: casting alloys and wrought alloys. Some alloys can be heat treated. Aluminium and its alloys have a low density, high electrical and thermal conductivity and excellent corrosion resistance. Tensile strength tends to be of the order of 150 to 400 MPa with the tensile modulus about 70 GPa. There is a high strength-to-weight ratio.

Chromium   Chromium is mainly used as an alloying element in stainless steels, heat-resistant alloys and high-strength alloy steels. It is generally used in these for the corrosion and oxidation resistance it confers on the alloys.

Cobalt   Cobalt is widely used as an alloy for magnets, typically 5-35% cobalt with 14-30% nickel and 6-13% aluminium. Cobalt is also used for alloys which have high strength and hardness at room and high temperatures. These are often referred to as Stellites. Cobalt is also used as an alloying element in steels.

Copper   Copper is very widely used in the commercially pure form and alloyed in the form of brasses, bronzes, cupro-nickels and nickel silvers. Brasses are copper–zinc alloys containing up to 43% zinc. Bronzes are copper–tin alloys. Copper–aluminium alloys are referred to as aluminium bronzes, copper–silicon alloys as silicon bronzes. Copper–beryllium alloys as beryllium bronzes. Cupro-nickels are copper–nickel alloys. Copper and its alloys have good corrosion resistance, high electrical and thermal conductivity, good machineability, can be joined by soldering, brazing and welding, and generally have good properties at low temperatures. The alloys have tensile strengths ranging from about 180 to 300 MPa and a tensile modulus about 20 to 28 GPa.

Gold   Gold is very ductile and readily cold worked. It has good electrical and thermal conductivity.

Iron   The term ferrous alloys is used for the alloys of iron. These alloys include carbon steels, cast irons, alloy steels and stainless steels. Steels have 0.05–2% carbon, cast irons 2–4.3% carbon. The term carbon steel is used for those steels in which essentially just iron and carbon are present. Steels with between 0.10% and 0.25% are termed mild steels, between 0.20% and 0.50% medium-carbon steels and 0.50–2% carbon as high-carbon steels. With such steels in the annealed state the tensile strength increases from about 250 MPa at low carbon content to 900 MPa at high carbon content, the higher the carbon content, the more brittle the alloy. The term low-alloy steel is used for alloy steels when the alloying additions are less than 2%, medium-alloy between 2% and 10% and high-alloy when over 10%. In all cases the carbon content is less that 1%. Examples of low-alloy steels are manganese steels with strengths of the order of 500 MPa in the annealed state and 700 MPa when quenched and tempered. Stainless steels are high-alloy steels with more than 12% chromium. The modulus of elasticity of steels tend to be about 200 to 207 GPa.

Lead   Other than its use in lead storage batteries, it finds a use in lead–tin alloys as a metal solder and in steels to improve the machinability.

Magnesium   Magnesium is used in engineering alloyed mainly with aluminium, zinc and manganese. The alloys have a very low density and though tensile strengths are only of the order of 250 MPa there is a high strength-to-weight ratio. The alloys have a low tensile modulus, about 40 GPa. They have good machinability.

Molybdenum   Molybdenum has a high density, high electrical and thermal conductivity and low thermal expansivity. At high temperatures it oxidizes. It is used for electrodes and support members in electronic tubes and light bulbs, and heating elements for furnaces. Molybdenum is, however, more widely used as an alloying element in steels. In tool steels it improves hardness, in stainless steels it improves corrosion resistance, and in general in steels it improves strength, toughness and wear resistance.

Nickel   Nickel is used as the base metal for a number of alloys with excellent corrosion resistance and strength at high temperatures. The alloys are basically nickel–copper and nickel–chromium–iron. The alloys have tensile strengths between about 350 and 1400 MPa, the tensile modulus being about 220 GPa.

Niobium   This has a high melting point, good oxidation resistance and low modulus of elasticity. Niobium alloys are used for high-temperature items in turbines and missiles. It is used as an alloying element in steels.

Palladium   This metal is highly resistant to corrosion. It is alloyed with gold, silver or copper to give metals which are used mainly for electrical contacts.

Platinum   This metal has a high resistance to corrosion, is very ductile and malleable, but expensive. It is widely used in jewellery. Alloyed with elements such as iridium and rhodium, the metal is used in instruments for items requiring a high resistance to corrosion.

Silver   Silver has a high thermal and electrical conductivity, and is very soft and ductile.

Tantalum   Tantalum is a high melting point, highly acid-resistant, very ductile metal. Tantalum–tungsten alloys have high melting points, high corrosion resistance and high tensile strength.

Tin   Tin has a low tensile strength, is fairly soft and can be very easily cut. Tin plate is steel plate coated with tin, the tin conferring good corrosion resistance. Solders are essentially tin alloyed with lead and sometimes antimony. Tin alloyed with copper and antimony gives a material widely used for bearing surfaces.

Titanium   Titanium as a commercially pure or alloy has a high strength coupled with a relatively low density. It retains its properties over a wide temperature range and has excellent corrosion resistance. Tensile strengths are typically of the order of 1100 MPa and tensile modulus about 110 GPa.

Tungsten   This is a dense metal with the highest melting point of any metal (3410°C). It is used for light bulb and electronic tube filaments, electrical contacts, and as an alloying element in steels. As whiskers it is used in many metal–whisker composites.

Zinc   Zinc has very good corrosion resistance and hence finds a use as a coating for steel, the product being called galvanized steel. It has a low melting point and hence zinc alloys are used for products such as small toys, cogs, shafts, door handles, etc. produced by die casting. Zinc alloys are generally about 96% zinc with 4% aluminium and small amounts of other elements or 95% zinc with 4% aluminium, 1% copper and small amounts of other elements. Such alloys have tensile strength of about 300 MPa, elongations of about 7-10% and hardness of about 90 BH.

Engineering polymers

The following is an alphabetical listing of the main polymers used in engineering, together with brief notes of their main characteristics.

Acrylonitrile–butadiene–styrene (ABS)   ABS is a thermoplastic polymer giving a range of opaque materials with good impact resistance, ductility and moderate tensile (17 to 58 MPa) and compressive strength. It has a reasonable tensile modulus (1.4 to 3.1 GPa) and hence stiffness, with good chemical resistance.

Acetals   Acetals, i.e. polyacetals, are thermoplastics with properties and applications similar to those of nylons. A high tensile strength (70 MPa) is retained in a wide range of environments. They have a high tensile modulus (3.6 GPa) and hence stiffness, high impact resistance and a low coefficient of friction. Ultraviolet radiation causes surface damage.

Acrylics   Acrylics are transparent thermoplastics, trade names for such materials including Perspex and Plexiglass. They have high tensile strength (50 to 70 MPa) and tensile modulus (2.7 to 3.5 GPa), hence stiffness, good impact resistance and chemical resistance, but a large thermal expansivity.

Butadiene–acrylonitrile This is an elastomer, generally referred to as nitrile or Buna-N rubber (NBR). It has excellent resistance to fuels and oils.

Butadiene–styrene This is an elastomer and is very widely used as a replacement for natural rubber because of its cheapness. It has good wear and weather resistance, good tensile strength, but poor resilience, poor fatigue strength and low resistance to fuels and oils.

Butyl Butyl, i.e. isobutene–isoprene copolymer, is an elastomer. It is extremely impermeable to gases.

Cellulosics This term encompasses cellulose acetate, cellulose acetate butyrate, cellulose acetate propionate, cellulose nitrate and ethyl cellulose. All are thermoplastics. Cellulose acctate is a transparent material. Additives are required to improve toughness and heat resistance. Cellulose acetate butyrate is similar to cellulose acetate but less temperature sensitive and with a greater impact strength. It has a tensile strength of 18 to 48 MPa and a tensile modulus of 0.5 to 1.4 GPa. Cellulose nitrate colours and becomes brittle on exposure to sunlight. It also burns rapidly. Ethyl cellulose is tough and has low flammability.

Chlorosulphonated polyethylene This is an elastomer, trade name Hypalon. It has excellent resistance to ozone with good chemical resistance, fatigue and impact properties.

Epoxies Epoxy resins are, when cured, thermosets. They are frequently used with glass fibres to form composites. Such composites have high strength, of the order of 200 to 420 MPa, and stiffness, about 21 to 25 GPa.

Ethylene propylene This is an elastomer. The copolymer form, EPM, and the terpolymer form, EPDM, have very high resistance to oxygen, ozone and heat.

Ethylene vinyl acetate This is an elastomer which has good flexibility, impact strength and electrical insulation properties.

Fluorocarbons These are polymers consisting of fluorine attached to carbon chains. See *Polytetrafluoroethylene*.

Fluorosilicones See *Silicone rubbers*.

Melamine formaldehyde The resin, a thermoset, is widely used for impregnating paper to form decorative panels, and as a laminate for table and kitchen unit surfaces. It is also used with fillers for moulding knobs, handles, etc. It has good chemical and water resistance, good colourability and good mechanical strength (55 to 85 MPa) and stiffness (7.0 to 10.5 GPa).

Natural rubber This is an elastomer. It is inferior to synthetic rubbers in oil and solvent resistance and oxidation resistance. It is attacked by ozone.

Nylons The term nylon is used for a range of thermoplastic materials having the chemical name of polyamides. A numbering system is used to distinguish between the various forms, the most common engineering ones being nylon 6, nylon 6.6 and nylon 11. Nylons are translucent materials with high tensile strength and of medium stiffness. Tensile strengths are typically about 75 MPa and the tensile modulus about 1.1 to 3.3 GPa. Additives such as glass fibres are used to increase strength. Nylons have low coefficients of friction, which can be further reduced by suitable additives. For this reason they are widely used for gears and rollers. All nylons absorb water.

Phenol formaldehyde   This is a thermoset and is mainly used as a reinforced moulding powder. It is low cost, and has good heat resistance, dimensional stability, and water resistance. Unfilled it has a tensile strength of 35 to 55 MPa and a tensile modulus of 5.2 to 7.0 GPa.

Polyacetal   See *Acetals*.

Polyamides   See *Nylons*.

Polycarbonates   Polycarbonates are transparent thermoplastics with high impact strength, high tensile strength (55 to 65 MPa), high dimensional stability and good chemical resistance. They are moderately stiff (2.1 to 2.4 GPa). They have good heat resistance and can be used at temperatures up to 120°C.

Polychloroprene   This, usually called neoprene, is an elastomer. It has good resistance to oils and good weathering resistance.

Polyesters   Two forms are possible, thermoplastics and thermosets. Thermoplastic polyesters have good dimensional stability, excellent electrical resistivity and are tough. They discolour when subject to ultraviolet radiation. Thermoset polyesters are generally used with glass fibres to form composite materials.

Polyethylene   Polyethylene, or polythene, is a thermoplastic material. There are two main types: low density (LDPE) which has a branched polymer chain structure and high density (HDPE) with linear chains. Materials composed of blends of the two forms are available. LDPE has a fairly low tensile strength (8 to 16 MPa) and tensile modulus (0.1 to 0.3 GPa), with HDPE being stronger (22 to 38 MPa) and stiffer (0.4 to 1.3 GPa). Both forms have good impermeability to gases and very low absorption rates for water.

Polyethylene terephthalate (PET)   This is a thermoplastic polyester. It has good strength (50 to 70 MPa) and stiffness (2.1 to 4.4 GPa), is transparent and has good impermeability to gases. It is widely used as bottles for fizzy drinks.

Polypropylene   Polypropylene is a thermoplastic material with a low density, reasonable tensile strength (30 to 40 MPa) and stiffness (1.1 to 1.6 GPa). Its properties are similar to those of polyethylene. Additives are used to modify the properties.

Polypropylene oxide   This is an elastomer with excellent impact and tear strengths, good resilience and good mechanical properties.

Polystyrene   Polystyrene is a transparent thermoplastic. It has moderate tensile strength (35 to 60 MPa), reasonable stiffness (2.5 to 4.1 GPa), but is fairly brittle and exposure to sunlight results in yellowing. It is attacked by many solvents. Toughened grades, produced by blending with rubber, have better impact properties. They have a strength of about 17 to 42 MPa and stiffness of 1.8 to 3.1 GPa.

Polysulphide   This is an elastomer with excellent resistance to oils and solvents, and low permeability to gases. It can, however, be attacked by microorganisms.

Polysulphone   This is a strong, comparatively stiff, thermoplastic which can be used to a comparatively high temperature. It has good dimensional stability and low creep with a strength of about 70 MPa and a stiffness of about 2.5 GPa.

Polytetrafluoroethylene (PTFE) PTFE is a tough and flexible thermoplastic which can be used over a very wide temperature range. Because other materials will not bond with it, the material is used as a coating for items where non-stick facilities are required.

Polyvinyls Polyvinyls are thermoplastics and include polyvinyl acetate, polyvinyl butyral, polyvinyl chloride (PVC), chlorinated polyvinyl chloride and vinyl copolymers. Polyvinyl acetate (PVA) is widely used in adhesives and paints. Polyvinyl butyral (PVB) is mainly used as a coating material or adhesive. PVC has high strength (52 to 58 MPa) and stiffness (2.4 to 3.1 GPa), being a rigid material. It is frequently combined with plasticizers to give a lower strength, less rigid, material. Chlorinated PVC is hard and rigid with excellent chemical and heat resistance. Vinyl copolymers can give a range of properties according to the constituents and their ratio. A common copolymer is vinyl chloride with vinyl acetate in the ratio 85 to 15. This is a rigid material. A more flexible form has the ratio 95 to 5.

Silicone rubbers Silicone rubbers or, as they are frequently called, fluorosilicone rubbers have good resistance to oils, fuels and solvents at high and low temperatures. They do, however, have poor abrasion resistance.

Styrene–butadiene–styrene This is called a thermoplastic rubber. Its properties are controlled by the ratio of styrene to butadiene. The properties are comparable to those of natural rubber.

Urea formaldehyde This is a thermosetting material and has similar applications to melamine formaldehyde. Surface hardness is very good. The resin is also used as an adhesive.

Engineering ceramics

The term ceramics covers a wide range of materials and here only a few of the more commonly used engineering ceramics are considered.

Alumina Alumina, i.e. aluminium oxide, is a ceramic which finds a wide variety of uses. It has excellent electrical insulation properties and resistance to hostile environments. Combined with silica it is used as refractory bricks.

Boron Boron fibres are used as reinforcement in composites with materials such as nickel.

Boron nitride This ceramic is used as an electric insulator.

Carbides A major use of ceramics is, when bonded with a metal binder to form a composite material, as cemented tips for tools. These are generally referred to as bonded carbides, the ceramics used being generally carbides of chromium, tantalum, titanium and tungsten.

Chromium carbide See *Carbides.*

Chromium oxide This ceramic is used as a wear-resistant coating.

Glasses The basic ingredient of most glasses is silica, a ceramic. Glasses tend to have low ductility, a tensile strength which is markedly affected by microscopic defects and surface scratches, low thermal expansivity and conductivity (and hence poor resistance to thermal shock), good resistance to chemicals and good electrical insulation properties. Glass fibres are frequently used in composites with polymeric materials.

Kaolinite This ceramic is a mixture of aluminium and silicon oxides, being a clay.

Magnesia Magnesia, i.e. magnesium oxide, is a ceramic and is used to produce a brick called a dolomite refractory.

Pyrex This is a heat-resistant glass, being made with silica, limestone and boric oxide. See *Glasses*.

Silica Silica forms the basis of a large variety of ceramics. It is, for example, combined with alumina to form refractory bricks and with magnesium ions to form asbestos. It is the basis of most glasses.

Silicon nitride This ceramic is used as the fibre in reinforced materials, such as epoxies.

Soda glass This is the common window glass, being made from a mixture of silica, limestone and soda ash. See *Glasses*.

Tantalum carbide See *Carbides*.

Titanium carbide See *Carbides*.

Tungsten carbide See *Carbides*.

Answers

The following are the numerical answers to problems and brief clues to the form of the answers for other problems.

Chapter 1

1   These might include (a) ease of forming in one piece, easily cleaned, stain resistant, waterproof; (b) stiff, strong, cheap; (c) leakproof, suitable for hot liquids, cheap, not easily broken; (d) good conductor, flexible; (e) cheap to make, wear resistant during handling, stiff; (f) withstands changing forces, stiff, strong, withstands impact forces; (g) attractive appearance, cheap to form

2   (a) Stainless steel; (b) wood; (c) china (a ceramic); (d) copper; (e) alloys of copper (cupronickel or bronze depending on the colour of the coins); (f) steel; (g) plastic, e.g. ABS

3   (a) Modulus; (b) ductility; percentage elongation; (c) fracture toughness; (d) strength; (e) electrical resistivity/conductivity; (f) thermal conductivity; (g) corrosive properties

4   Strong and brittle

5   Strong and tough

6   20 MPa

7   0.67%

8   50 kN

9   12%

10 50 kN

11 The bronze is stronger and more ductile

12 Stronger in compression, brittle

13 Ductile above 0°C, brittle below

14 Reasonably good

15 Very low resistivity, of the order of $10^{-8}\ \Omega$ m

16 0.0125 Ω

17 Thermoplastics: flexible, soft, can be formed by heating; thermosets: rigid, hard, cannot be formed by heating

18 Brittle, must not be subject to sudden forces or sudden changes in temperature

19 128 MPa/Mg m^{-3}, 33 MPa/Mg m^{-3}, 0.78 £/MPa, 6 £/MPa

Chapter 2

1 (a) 420 MPa, (b) 62%, (c) 18-40%, (d) 355 MPa, (e) 510 MPa, (f) 1020-1070 kg/m^3, (g) 3.0-4.5 MPa m$^{-1/2}$, (h) 20 MPa, (i) 11-13 × 10^{-2} K^{-1}, (j) 1.4-3.1 GPa

2 Cast iron 0.014 GPa/kg m^{-3}, Al alloy 0.027 GPa/kg m^{-3}, PVC 0.002 GPa/kg m^{-3}

3 Steel 220 GPa, Al alloy 71 GPa, Polypropylene 1-2 GPa, Composite 20 GPa

4 120 MPa/Mg m^{-3}

5 1.4-3.1 GPa, in the high range of modulus values for plastics

6 470-570 MPa, 170-280 MPa, 18-35%

7 150M36: 620-770 MPa, 400 MPa, 18%; 530M40: 700-850 MPa, 525 MPa, 17%

8 LM6

9 Polyacetal

10 226M44

Chapter 3

1 (a) 61 GPa, (b) 380 MPa

2 (a) 10.8 MPa, (b) 1.1 GPa

3 (a) 660 MPa, (b) 425 MPa, (c) 200 GPa

4 31.1%

5 (a) 480 MPa, (b) 167 GPa

6 300 MPa, 280 MPa

7 2.5 GPa, 80 MPa

8 See Figure A.1

9 Stronger and less ductile

10 (a) Titanium alloy, (b) nickel alloy

11 Cellulose acetate

12 Becoming more ductile

13 Becoming more brittle

14 As the temperature drops becoming more brittle

15 Becoming more ductile

16 HV 198

17 HV 275

18 HV 71

19 HB 217

20 HB 57

21 After exposure breakdown voltage decreases

22 Ni–Cr alloy most corrosion resistant

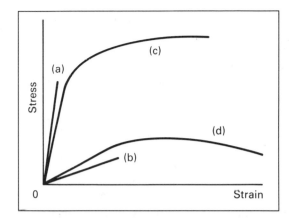

Figure A.1   *Chapter 3 Problem 8*

23 Increasing carbon reduces oxidation, increasing chromium reduces oxidation
24 Industrial pollutants more damaging than marine conditions, with rural surroundings being least corrosive
25 (a) Hardness test, (b) impact test, (c) tensile test for the modulus of elasticity, (d) impact, or tensile or hardness, test, (e) bend test

Chapter 4

1 A crystal within a metal, i.e. a region of orderly packing of atoms
2 A mixture of two or more elements, e.g. iron and carbon in steel
3 Ferrous alloys have iron as the main constituent, a non-ferrous alloy a metal other than iron
4 An array of grains, i.e. crystals, within which there are orderly arrays of atoms
5 More ductile, the bigger the grains
6 Elongated grains give different properties in the directions of the grains compared with at right angles
7 Grains become elongated and distorted with an increasing number of dislocations. The tensile strength and hardness increases, the ductility decreases
8 See Figure 4.7
9 (a) Large grain, few dislocations, (b) small grain, many dislocations
10 Increase in dislocations and hence an increase in yield strength, tensile strength and hardness but a decrease in ductility
11 See Figure 4.19
12 See Figure 4.19. Above the recrystallization temperature no work hardening occurs
13 Grains elongated in direction of rolling

14 Cold rolled has distorted grains, is work hardened and directionality of properties. Hot rolled has large grains, is ductile and has surface oxide layers

15 Greater than 400°C, probably about 500°C

16 (a) More distortion and dislocations, increased hardness and brittleness, (b) up to 300°C, (c) above 300°C

17 (a) About 110 HV, (b) about 30 HV, (c) roll, anneal, roll, anneal, roll so that the final rolling gives less than 10% reduction

18 See Figures 4.22 and 4.23. The greater the crystallinity, the greater the density, melting point and strength

19 LDPE is a branched polymer with less crystallization than HDPE which is a linear polymer. See Table 4.1

20 (a) To protect against UV and resist deterioration, (b) to make more flexible, (c) to reduce cost, increase perhaps strength, impact strength, resistivity, or reduce friction

21 See Table 4.3

22 Makes it more rigid

23 See Figure 4.31 and associated text

24 36.4 GPa, in direction of fibres

25 182.2 GPa

26 205 GPa

27 The long fibres give directionality of properties and a greater improvement in strength and modulus than random fibres, which give no directionality

Chapter 5

1 Better surface finish with cold drawing. The heating anneals the material to make it soft and ductile

2 Oriented distorted grains and hence work hardened with directionality of properties

3 Die casting

4 Slow cooling gives high degree of crystallinity

5 Molecules aligned along the direction of the extrusion

6 Molecules aligned with the direction in which the material was stretched

7 Molecules lined up along the length of the bag

8 Gives molecular alignment and improves the strength

9 (a) Recrystallization and grain growth, ductility improves; (b) martensite forms as carbon atoms become trapped, increase in hardness; (c) some carbon atoms diffuse out of martensite, increase in ductility; (d) fine particles slowly move out of quenched material into dislocations and grain boundaries, increase in hardness; (e) surface layers become martensitic, increase in hardness, (f) carbon diffuses into outer layers, increase in surface hardness

10 Annealing gives grain growth and a soft structure; quenching gives martensitic structure and increase in hardness, strength and brittleness; tempering allowing carbon atoms to diffuse out of the martensite and so reduce the structural distortion and hence brittleness

11 Annealing gives grain growth and a soft, weak structure; work hardening distorts grains and introduces dislocations with the result that the material is stronger, harder and more brittle

12 Annealing gives grain growth and a soft, weak structure, precipitation hardening causes fine particles to become lodged at grain boundaries and dislocations, hence increasing the strength, hardness and brittleness

13 The hammer head is forged and then the striking surface is surface hardened, possibly by flame hardening followed by tempering

14 The blade has to be tough with the teeth hard. Cast ingots are hot rolled, then blade-size strips cut out and teeth machined. Surface hardening, flame hardening, is then used for the teeth followed by tempering in order to achieve the required hardness and not too brittle a state for the teeth

Chapter 6

1 (a) Aluminium alloy, e.g. LM9, die cast; (b) carbon steel, about 0.3% carbon, or low-alloy steel such as a chromium–molybdenum steel, hot rolling; (c) unplasticized PVC, extrusion; (d) high-density polyethylene, injection moulding; (e) high-density polyethylene, extrusion; (f) nylon, injection moulding or die cast aluminium alloy, e.g. LM9; (g) acrylic (polymethyl methacrylate), injection moulding; (h) medium-carbon steel, about 0.4% carbon, or low-alloy steel such as 530M40, forged, quenched and tempered; (i) polypropylene, injection moulded or aluminium alloy, e.g. LM6, die cast; (j) ABS, injection moulding

2 (a) Thermoset, e.g. urea formaldehyde, moulding; (b) medium-carbon steel or stainless steel, forging; (c) copper, e.g. C106, extrusion; (d) ABS, injection moulding

3 See the British Standards

Chapter 7

1 See Sections 7.1.1 and 7.1.2

2 (a) Improvements are required, (b) the process or equipment is prohibited from being used

3 See the HSE booklets

4 (a) Contravenes HSW Act duties imposed on employees, (b) employers are failing to ensure the welfare of employees in an accident or emergency, (c) contravenes HSW Act duties imposed on employees, (d) the inspector has the right, (e) the worker cannot refuse, (f) should report it

5 Could be: (a) spectacles, goggles, face screens; (b) earplugs or muffs; (c) safety boots, gloves

Index

Index

Index